Palgrave Studies in Law, Neuroscience, and Human Behavior

Neuroscience is drawing increasing attention from lawyers, judges, and policy-makers because it both illuminates and questions the myriad assumptions that law makes about human thought and behavior. Additionally, the technologies used in neuroscience may provide lawyers with new forms of evidence that arguably require regulation. Thus, both the technology and applications of neuroscience involve serious questions implicating the fields of ethics, law, science, and policy. Simultaneously, developments in empirical psychology are shedding scientific light on the patterns of human thought and behavior that are implicated in the legal system. The Palgrave Series on Law, Neuroscience, and Human Behavior provides a platform for these emerging areas of scholarship.

More information about this series at
http://www.palgrave.com/gp/series/15605

Sjors Ligthart · Dave van Toor ·
Tijs Kooijmans · Thomas Douglas ·
Gerben Meynen
Editors

Neurolaw

Advances in Neuroscience, Justice & Security

palgrave
macmillan

Editors
Sjors Ligthart
Criminal Law
Tilburg University
Tilburg, The Netherlands

Dave van Toor
Criminal Law
Utrecht University
Utrecht, The Netherlands

Tijs Kooijmans
Criminal Law
Tilburg University
Tilburg, The Netherlands

Thomas Douglas
Philosophy
University of Oxford
Oxford, UK

Gerben Meynen
Forensic Psychiatry
Utrecht University
Utrecht, The Netherlands

Ethics and psychiatry
VU Amsterdam
Amsterdam, The Netherlands

Palgrave Studies in Law, Neuroscience, and Human Behavior
ISBN 978-3-030-69276-6 ISBN 978-3-030-69277-3 (eBook)
https://doi.org/10.1007/978-3-030-69277-3

Cover illustration: traffic_analyzer, Getty Images

This Palgrave Macmillan imprint is published by the registered company Springer Nature Switzerland AG
The registered company address is: Gewerbestrasse 11, 6330 Cham, Switzerland

Preface

Neurolaw is a relatively young domain of interdisciplinary research on the promises and perils of neuroscience for the law, often focusing on criminal law. It covers a diversity of topics and approaches, some more theoretical—e.g. regarding the foundations of punishment—others more practical—e.g. concerning the use of brain scans in the courtroom.[1] A central question for neurolaw is how neuroscience could contribute to justice and security.

This book aims to contribute to the ongoing debate on neuroscience, justice, and security, by examining how neuroscience could contribute to fair and more effective criminal justice systems, and how both neuroscientific insights and information can be integrated into criminal law in a way that respects fundamental rights and moral values. The first part approaches these questions from a legal perspective, followed by ethical accounts in part two.

[1] See, e.g. Vincent et al. (2020), Ryberg (2020), Bigenwald and Chambon (2019), Foquaert et al. (2020), Meynen (2014, 2020), Ligthart et al. (2020), Bublitz (2020), Mecacci and Haselager (2019), Birks and Douglas (2018), Ienca and Andorno (2017), Kellmeyer et al. (2016), Pardo and Patterson (2015), Morse and Roskies (2013), Shen (2013), Farahany (2012), Simpson (2012), Richmond et al. (2012), Greely and Wagner (2011), and Green and Cohen (2004).

In the first part, David Linden starts with discussing the possibilities and limitations of neuroscience in criminal legal proceedings. More specifically, he considers whether and how neuroscience could contribute to assessing individual mental states and to answering central questions of criminal law on action responsibility, *mens rea*, capacity, liability, re-offending, and prevention. Linden approaches these questions from both a retributivist as well as a consequentialist perspective on criminal law and punishment.

Next, Georgia Gkotsi discusses the interpretation of neuroscientific results by judges in criminal proceedings. She presents preliminary results of a focus group study, examining how judges would consider neuroscientific data in assessing an individual's risk of future dangerousness. Since overestimation of the importance of neurobiological data for the prediction of criminal behaviour is clearly a risk, Gkotsi argues that judges should be trained and informed about the limitations and the interpretative nature of neuroscientific data in relation to legal notions.

In the following chapter, Paul Catley considers the law of England and Wales and he makes a case for recognising an intermediate level of criminal responsibility between those deemed not criminally responsible and those held to be criminally responsible. This intermediate level would apply where individuals have a significantly impaired ability to conform their behaviour to what criminal law requires. In his chapter, Catley analyses how cognitive sciences could be helpful in framing such a partial defence of diminished capacity.

Subsequently, Lisa Claydon discusses how insights from cognitive sciences could be helpful to better understand when individuals should or should not be punished in the context of criminal law. Focussing on the law of England and Wales, she considers the criminal culpability of those who are coerced in committing a crime. She examines whether the law should recognize a criminal defence tailored to coercion and control, putting emphasis on the potential role of cognitive sciences in this regard.

Turning the perspective towards the empirical and normative limitations of applying neurotechnologies in criminal justice, Ewout Meijer and Dave van Toor consider the possibility of identifying memories in the brains of sleeping suspects. Whereas Meijer argues that memory detection in sleeping participants would, at least in theory, be possible from an empirical perspective, Van Toor contends that such employment of neurotechnological memory detection would infringe the privilege against self-incrimination under European human rights law.

Bridging the first and second part of this volume, Sjors Ligthart, Tijs Kooijmans, and Gerben Meynen argue that ethics and the law could learn from each other when analysing the normative boundaries of employing brain-reading technology in criminal justice and forensic psychiatry. In view of facilitating an integrative legal-ethical approach on this issue, the authors identify three central ethical values—autonomy, confidentiality, and trust—and explore whether and how they are reflected in the legal debate.

In the second part, combining law and ethics, Cristina Scarpazza, Colleen Berryessa, and Farah Focquaert discuss ethical and legal implications of the medical distinctions between criminal offenders suffering from either idiopathic or acquired paedophilia. Based on the current scientific knowledge regarding both disorders, the authors argue that retributive punishments are unlikely to tackle the problems related to future paedophilic behaviour. Instead, alternative strategies may be needed to prevent future offending by individuals with both idiopathic and acquired paedophilia.

Next, Thomas Douglas and Lisa Forsberg examine the moral rationale of recognising a legal right to mental integrity. In view of emerging neurotechnologies that enable to enter and alter peoples' minds, it has been argued that the law should introduce a right against (certain kinds of) non-consensual interference with the mind, i.e. a right to mental integrity. However, as yet, the arguments for its recognition remain unclear. Douglas and Forsberg seek to make some progress towards a systematic account of the rationales for a right to mental integrity, focusing on three distinctive appeals: the appeal to intuition, to justificatory consistency, and to technological development.

Jesper Ryberg continues the debate on deploying neurointerventions in crime prevention. He emphasises the importance of how we interpret the question of ethical legitimacy of administering neurointerventions in criminal justice: either as asking whether it can *ever* be justified to use neurointerventions in a particular way to prevent recidivism, or, alternatively, whether it would be justified to use neurointerventions within the criminal justice context that *currently* exists (or will exist in the near future). Ryberg argues that these two ways of understanding the question on ethical legitimacy may lead to very different answers.

Considering the impact of neuroscience for our understanding of punishment and criminal law, Bebhinn Donnelly-Lazarov explores whether persons with an extremely 'good' brain are morally better or worse than the rest of us. In doing so, Bebhinn Donnelly-Lazarov discusses some ethical implications of neuroenhancement for current approaches of criminal offending.

Finally, Andrea Lavazza and Flavia Corso argue that both neuroscientific insights and the employment of neurotechnologies in criminal practice will not necessarily conflict with retributivist intuitions. As the authors contend, by adopting a naturalisation approach of criminal law and punishment, both the consequentialist model and the retributive account can be plausibly naturalized and defended within a scientifically informed theory of punishment and criminal justice.

Altogether, these chapters provide a profound and diverse discussion of the possible implications of neuroscience for the criminal justice system. They illustrate the thoroughly interdisciplinary nature of the debate, in which science, law, and ethics are closely intertwined. We hope that the essays in this volume help to find valuable ways forward for neuroscience, justice, and security.

Tilburg, The Netherlands — Sjors Ligthart
Utrecht, The Netherlands — Dave van Toor
Tilburg, The Netherlands — Tijs Kooijmans
Oxford, UK — Thomas Douglas
Utrecht, The Netherlands — Gerben Meynen

References

Bigenwald, A., & Chambon, V. (2019). Criminal responsibility and neuroscience: No revolution yet. *Frontiers in Psychology, 10*.

Birks, D., & Douglas, T. (Eds.). (2018). *Treatment for crime: Philosophical essays on neurointerventions in criminal justice*. Oxford University Press.

Bublitz, J. C. (2020). The nascent right to psychological integrity and mental self-determination. In A. Von Arnauld, K. Von der Decken, & M. Susi (Eds.), *The Cambridge handbook of new human rights: Recognition, novelty, rhetoric*. Padstow: Cambridge University Press.

Farahany, N. A. (2012). Incriminating thoughts. *Stanford Law Review, 64*, 351–408.

Focquaert, F., Caruso, G., Shaw, E., & Pereboom, D. (2020). Justice without retribution: Interdisciplinary perspectives, stakeholder views and practical implications. *Neuroethics, 13*, 1–3.

Green, J. & Cohen, J. (2004). For the law, neuroscience changes nothing and everything. *Philosophical Transactions of the Royal Society of London. Series B, Biological Sciences, 359*(1451), 1775–1785.

Greely, H. T., & Wagner, A. D. (2011). *Reference guide on neuroscience*. Washington, DC: National Academies Press/Federal Judicial Center.

Ienca, M., & Andorno, R. (2017). Towards new human rights in the age of neuroscience and neurotechnology. *Life Sciences, Society and Policy, 13*(5), 1–27.

Kellmeyer, P., et al. (2016). Effects of closed-loop medical devices on the autonomy and accountability of persons and systems. *Cambridge Quarterly of Healthcare Ethics, 25*, 623–633.

Ligthart, S., Douglas, T., Bublitz, C., Kooijmans, T., & Meynen, G. (2020). Forensic brain-reading and mental privacy in European human rights law: Foundations and challenges. *Neuroethics*. https://doi.org/10.1007/s12152-020-09438-4.

Mecacci, G., & Haselager, P. (2019). Identifying criteria for the evaluation of the implications of brain reading for mental privacy. *Science and Engineering Ethics, 25*, 443–461.

Meynen, G. (2014). Neurolaw: Neuroscience, ethics, and law. Review essay. *Ethical Theory and Moral Practice, 17*, 819–829.

Meynen, G. (2020). Neuroscience-based psychiatric assessments of criminal responsibility: Beyond self-report? *Cambridge Quarterly of Healthcare Ethics, 29*, 446–458.

Morse, S. J., & Roskies, A. L. (Eds.). (2013). *A primer on criminal law and neuroscience*. New Yok: Oxford University Press.

Pardo, S., & Patterson, D. (2015). *Minds, brains, and law. The conceptual foundations of law and neuroscience*. Oxford University Press.

Richmond, S., Rees, G., & Edwards, S. J. L. (Eds.). (2012). *I know what you're thinking*. Oxford University Press.

Ryberg, J. (2020). *Neurointerventions, crime, and punishment. Ethical considerations*. New York: Oxford University Press.

Shen, F. X. (2013). Neuroscience, mental privacy and the law. *Harvard Journal of Law & Public Policy, 36*, 653–713.

Simpson, J. R. (Ed.). (2012). *Neuroimaging in forensic psychiatry: From the clinic to the courtroom*. Wiley-Blackwell.

Vincent, N. A., Nadelhoffer, T., & McCay, A. (2020). *Neurointerventions and the law: Regulating human mental capacity*. New York: Oxford University Press.

Contents

Notes on Contributors

Dr. Colleen Berryessa is assistant professor at the school of a criminal justice at Rutgers university. In her research, utilizing both qualitative and quantitative methods, she considers how different psychological processes, perceptions, attitudes, and social contexts affect the criminal justice system, particularly related to courts and sentencing. She primarily examines these issues, using both social psychological and socio-legal lenses, in relation to two areas: (1) how these phenomena affect the discretion of criminal justice actors in their responses to offending and decision-making in courts; (2) how these phenomena affect lay views and consideration of courts, sentencing systems, and punishment practices.

Paul Catley is Professor of Neurolaw and until April 2021 was Head of the Open University Law School. His research focuses on the use and potential use of neuroscientific and genetic evidence in the courts and within justice systems more widely. His interests are wide ranging and include the use of neuroscientific evidence to detect memory and lies, the use of brain scanning to inform treatment and end of life decisions for patients with persistent disorders of consciousness and the appropriate approaches of the law in cases where brain impairment or brain injury may affect responsibility and/or capacity.

Dr. Lisa Claydon examines criminal law, with a particular interest in mental condition and other defences that are based on excusing conditions. She is actively researching the intersection between cognitive neuroscience and the criminal law. She was co-investigator on an AHRC-funded project entitled *A Sense of Agency*. This project examined neurocognitive and legal approaches to a personal sense of agency. Currently, she is researching what neuroscience may tell us about memory in the courtroom and looking at the effect of alcohol and drugs on criminal responsibility.

Dr. Flavia Corso holds a II level master degree from the University of Genoa.

She collaborates with Andrea Lavazza on a variety of neuroethical issues, mainly focusing on the role of neuroscience in the field of criminal justice. Recently, she contributed to an Italian volume on neuroethics, discussing the neuro-legal issue of responsibility in the age of neuroscience.

Bebhinn Donnelly-Lazarov As Professor of Neuroscience, Law and Legal Philosophy, Bebhinn's research interests lie in jurisprudence and criminal law theory. Her book on criminal attempts, published by Cambridge University Press in 2015, is structured around an Anscombian account of intentional action. Recent and ongoing work explores our understanding of the mind and considers its implications for criminal responsibility, *mens rea*, and for defences. Bebhinn has begun to write a book on law and consciousness.

Thomas Douglas is Professor of Applied Philosophy at the Oxford Uehiro Centre for Practical Ethics, where he is Director of Research and Development. He is also a Senior Research Fellow at Jesus College, Editor of the Journal of Practical Ethics, and Principal Investigator on the project 'Protecting Minds: The Right to Mental Integrity and the Ethics of Arational Influence', funded by a Consolidator Award from the European Research Council. His research lies mainly in practical and normative ethics and currently focuses on the ethics of predicting and influencing behaviour.

Dr. Farah Focquaert is Professor of philosophical anthropology at the Department of Philosophy and Moral Sciences, Ghent University, affiliated with the Bioethics Institute Ghent and Co-director of the International Justice Without Retribution Network. Her research interest lies in the philosophy of free will, responsibility and punishment, and in the field of neuroethics. She is the first editor of the Routledge Handbook of the Philosophy and Science of Punishment (2021).

Dr. Lisa Forsberg is a British Academy Postdoctoral Fellow in the Faculty of Law, and (in Philosophy) at Somerville College and the Oxford Uehiro Centre for Practical Ethics. Her main research interests lie in normative and practical ethics, and in the philosophy of medical and criminal law. Her postdoctoral project, 'Changing One's Mind: Neurointerventions, Autonomy, and the Law on Consent', is on medical consent and examines the extent to which English law on consent sufficiently protects morally salient patient interests.

Dr. Georgia Gkotsi is a Research Fellow at the Faculty of Law of the University of Athens, Greece. After receiving a law degree from the University of Athens, she completed a Master's in Philosophy of Law and Bioethics at the same University, followed by a Master's in Comparative Law at the Universite Paris 1 - Pantheon Sorbonne. She received her PhD from the University of Lausanne, Switzerland. Her dissertation dealt with the ethical and legal implications of the use of neuroimaging techniques in criminal courts. Her research expertise lies in the area of mental health law, human rights of mentally disabled persons, neurolaw, and bioethics.

Tijs Kooijmans is a full Professor of Criminal Law at Tilburg University and a substitute judge at the 's-Hertogenbosch Court of Appeals. He is a co-author of a leading handbook about Dutch criminal procedure and a commentator of Dutch leading criminal cases. His research interest lies in criminal law in general, confiscation and seizure of illegally obtained assets, forensic psychiatry, and neurolaw.

Dr. Andrea Lavazza is a Senior Research Fellow in Neuroethics at Centro Universitario Internazionale, and adjunct Professor at the University of Pavia. He specializes in philosophy of mind and neuroethics. His main research is in the field of neuroethics. He has published papers on enhancement, memory manipulation, cognitive freedom, and human brain organoids. His interests are focused on moral philosophy, free will, and law at the intersection with cognitive sciences. He is working on naturalism and its relations with other kinds of causation and explanation in philosophy of mind and philosophical anthropology.

Sjors Ligthart holds a Master's in Criminal Law. He is currently completing his PhD thesis on coercive brain-reading in criminal law. Other research interests include the introduction of neurointerventions and virtual reality systems into the domain of criminal justice, the debate on fundamental neurorights, and the concept of legal insanity. He is a lecturer of Penitentiary Law and editorial secretary of the leading Dutch journal on criminal law.

David Linden is full Professor of Translational Neuroscience and Scientific Director of the School for Mental Health and Neuroscience. He is a Psychiatrist at Maastricht University Medical Centre. His specialist clinical areas include neuropsychiatry, genetic syndromes in psychiatry, mood disorders, psychosis, and alcohol dependence. His research focuses on mechanisms and treatment of mental and neurodegenerative disorders. His group combines neuroimaging, cognitive neuroscience, genetics, and clinical research in order to develop new biological models and find new treatment targets.

Dr. Ewout Meijer is an Assistant Professor at the Faculty of Psychology and Neuroscience. He obtained his PhD in 2008 with a dissertation on the use of psychophysiological measures in lie and memory detection. He has published about a variety of topics, including deception detection, investigative interviewing, and cheating behaviour. He served as a Research Fellow at the Hebrew University of Jerusalem in 2011–2012, and as a fellow of the Israel Institute for Advanced Studies in 2020–2021.

Gerben Meynen is Psychiatrist and Professor of Forensic Psychiatry, Willem Pompe Institute for Criminal Law and Criminology (Utrecht Centre for Accountability and Liability Law, UCALL), Utrecht University, and professor of Ethics and Psychiatry, Department of Philosophy, Humanities, Vrije Universiteit Amsterdam. His research interests include legal insanity and the implications of neuroscience for criminal law and forensic psychiatry.

Jesper Ryberg is Professor of Ethics and Philosophy of Law at the Department of Philosophy. He writes and teaches in the areas of ethics and philosophy of law. He is the head of the Research Group for Criminal Justice Ethics and is currently also head of the Neuroethics and Criminal Justice research project. Ryberg has published in philosophical journals such as The Philosophical Quarterly, Philosophical Papers, Theoria, Ethical Theory and Moral Practice, The Journal of Ethics, Res Publica, Journal of Medical Ethics, Neuroethics, Journal of Applied Philosophy, Social Theory and Practice, International Journal of Applied Philosophy, Criminal Law and Philosophy, Analysis, Utilitas, Ratio, and AJOB Neuroscience.

Dr. Cristina Scarpazza is Assistant Professor at the Department of General Psychology, University of Padova. Her research interest lies in psychology and neuroscience, with particular emphasis on early diagnosis of psychiatric disorders, identification of neuroanatomical signature of psychiatric illness, group to individual inferences, and forensic psychiatry (with particular focus on insanity evaluation). She is particularly interested in cognitive biases and their impact in the interpretation of scientific findings. Through her long-standing collaboration with the King's College London, she was actively involved in different projects aiming to improve the translational impact of neuroimaging findings from research to clinical practice.

Dr. Dave van Toor joined the Department of Criminal of the Radboud University Nijmegen in 2008. He started at the Universität Bielefeld in September 2014, first as a Research Assistant, later as Research Associate at the Department of Criminal (Procedural) Law & Criminology. In the meantime, he completed his doctoral dissertation on the legal implications of using coerced neuroscientific memory detection in criminal cases. Currently, he is Assistant Professor in Criminal (Procedural) Law at Utrecht University. His main research interest lies in the legitimacy of police investigations in view of human rights.

Legal Perspectives

Possibilities and Limitations of Neuroscience in the Legal Process

David Linden

Introduction

The neurosciences (broadly defined as comprising clinical and basic neuroscience and the science and clinical practice of mental health) are relevant to law and legal practice in two main ways. They contribute to jurisprudence by providing insight into the causes and mechanisms underlying human action (and people's perceptions about them) and they can contribute to individual cases with information about reliability of evidence, responsibility and dangerousness of the perpetrator (Jones et al., 2013). The first contribution has been discussed in detail by Greene and Cohen in an essay with the programmatic title "For the law, neuroscience changes nothing and everything"(Greene & Cohen, 2004). The authors argue that the current retributivist penal system, with the

D. Linden (✉)
Faculty of Health, Medicine and Life Sciences, School for Mental Health and Neuroscience, Maastricht University, Maastricht, The Netherlands
e-mail: david.linden@maastrichtuniversity.nl

S. Ligthart et al. (eds.), *Neurolaw*, Palgrave Studies in Law, Neuroscience, and Human Behavior,
https://doi.org/10.1007/978-3-030-69277-3_1

underlying assumption of free will, rests on dubious philosophical foundations. At least, it could be easily challenged by those who infer from the prevailing (largely) deterministic model of the universe that free will in the sense required for a strong concept of responsibility or blameworthiness is an illusion, in other words that free will is incompatible with the laws of physics (incompatibilism). They deny that abandoning classical concepts of blameworthiness would upend the penal system because all it would entail is to abandon retributivism and to base punishment completely on consequentialist goals (which already play a major role in the current system). They also point out that none of these considerations depend particularly on advances in neurosciences—natural and moral philosophy operate quite independently from them—but that these advances may influence folk psychology by providing more insight into the neurophysiological causal chains that led to the offensive action. Thus, the more people are informed about the incremental psychological stressors and resulting brain changes that preceded delinquency, the more they may be inclined to accept that a person is not to blame for his or her actions. In principle, though, it should not matter how much we find out about this causal chain because, ultimately, it exists for any human action, good or bad. Regardless of this increasing influence of neuroscience on folk psychology (and the resulting increasing willingness to exculpate perpetrators on the basis of the condition of their brains), it will still take some time and an extended debate until our legal systems will be aligned with the scientific quest for causal (mechanistic) explanation.

In the meantime, neuroscience can play a role in its second domain, supporting the legal process in the areas of evidence, assessment of the mental state of the offender at the time of the offence, disposal, and prognosis. In order to organize my discussion of these areas, I use the taxonomy of responsibility proposed by Vincent (Vincent, 2010). I first discuss the contribution of neuroscience to the gathering of evidence in the context of "action responsibility" (*actus reus*). I then consider the assessment of the offender's mental state under the headings of "intent" and "capacity". The contribution of neuroscience to the determination of disposal and prognosis will be discussed under the headings "liability responsibility" and "prevention of re-offending". Other authors of this book will expand on several of these topics, and I will provide cross-references to the relevant chapters. The advances of neuroscience, particularly functional neuroimaging, in the last 25 years have given rise to hopes that many questions of criminal evidence could soon be resolved by brain scans, for example for lie detection, and that quantitative imaging would provide insights into a person's mind that could be used for assessment of capacity responsibility and prognosis. However, partly because of the difficulty of validating such predictive algorithms, this hope has so far not been fulfilled (and Chapters 5 and 6 of this volume discuss whether such a future would even be desirable and compatible with human rights concepts). Conversely, classical individual psychometric and psychopathological assessments (sometimes supported by clinical neuroimaging) are still very much needed in the criminal court, and I will argue that this will still be the case when (as I assume) the legal systems will move to more consequentialist frameworks as proposed by Green and Cohen over the coming decades.

Action Responsibility

A primary aim of criminal proceedings is to determine the circumstances of the action that led to the outcome in question, for example injury to someone's body. This determination, which includes the identification of the perpetrator or perpetrators of the action, concerns what in classical legal terminology is called *actus reus*, in distinction from the determination of the *mens rea*, which concerns intention and capacity of the perpetrator. The determination of the *actus reus* is the key component of the investigative part of criminal proceedings and includes weighing the evidence provided by witnesses, suspects, and victims. Neuroscience methods have been suggested to be potentially useful for the assessment of the reliability of a source of evidence. This could include techniques for lie detection (evaluating whether a particular statement is made with the intent to deceive), identifying perpetrator's knowledge or determining general unreliability of a witness. These techniques are essentially extensions of classical psychological or psychophysiological techniques. Lie detection uses presumed physiological signatures of deception (mainly altered arousal levels) to evaluate whether a particular statement is likely to be true. Its reliability as an investigative method is a matter for ongoing debate. Theoretically, a lie detector could also be based on the recording of brain signatures of deception from EEG or functional MRI signals, but such attempts have to be regarded as premature given the current state of these fields (Rusconi & Mitchener-Nissen, 2013; Farah et al., 2014). The identification of perpetrator's (or "guilty") knowledge could use brain signature of familiarity, for example, with a crime scene. It is discussed further in Chapter 5 by Meijer and Van Toor. Finally, the general reliability of a witness depends on his or her cognitive abilities and personality profile. Its assessment is mainly within the domain of psychology (e.g. with tests of short- and long-term memory) or psychopathology (e.g. with regard to pathological lying), but brain imaging can help if there is a question of an identifiable brain disease (such as Alzheimer's dementia) that could make a witness unreliable.

Mens Rea

Determining a person's intention on the basis of brain imaging signals or other neural measures could be useful both for the evaluation of the truthfulness of a statement (in the context of "lie detection", as outlined in the previous section) and for the determination of the *mens rea*. Although it is possible to map correlates of people's mental states with functional imaging, for example detect brain activation patterns associated with auditory hallucinations (Linden, 2012), the reverse inference (from a brain state onto a mental state) is very difficult to make on the basis of brain imaging data (Poldrack, 2006). Many external stimuli, cognitive tasks, and presumably also mental states can be associated with similar or highly overlapping brain activation patterns. Thus, although we can predict reasonably well which brain areas will be involved in the processing of a particular stimulus, for example a face, we cannot infer from activation of the corresponding brain area (the fusiform face area) that the person was actually seeing a face—they could also have imagined or hallucinated a face or been exposed to visual stimuli that had some face-like features. More fine-grained inferences from brain states onto mental states may be possible through multivariate pattern analysis.

Here the general procedure entails training an algorithm on the brain activation patterns associated with events from two different categories (e.g. pictures of appetizing and non-appetizing food) and then testing how well it predicts what type of food a person was seeing during a new instance of presentation of food pictures (Franssen et al., 2020). An interesting attempt in this respect was an experiment modelling the culpable states "knowing" and "reckless" in the meaning of the American Model Penal Code, which provided preliminary evidence for the possibility of differentiating such states at the neural level (Vilares et al., 2017). This approach can be extended to more than two categories. One example is the differentiation of brain activation patterns associated with six basic emotions (disgust, fear, happiness, sadness, anger, and surprise) (Saarimäki et al., 2016). Such fMRI-based mapping of neural patterns associated with specific mental states relies heavily on the cooperation of the participant and although significant accuracy rates have been reported in many studies these are often just above chance level and thus not in the range needed for criminal evidence (Uncapher et al., 2015). At present, the associations between neuroimaging or psychophysiological markers and specific patterns of thought or behaviour are not yet stable enough for any prospect of replacing the classical clinical and personality assessments used to determine intent (and capacity responsibility, see the following section) in forensic psychology and psychiatry.

Capacity Responsibility

A person who committed a violent offence might have the defence of insanity available to them, if, in general, at the time of the offence they were suffering from a mental illness that precluded them from understanding the nature or wrongfulness of their action or, if they had such understanding, from acting upon it (Simon & Ahn-Redding, 2006). The first scenario, also called "cognitive insanity" may occur, for example, in patients with dementia or delusions and is available in most jurisdictions; the latter, also called "volitional insanity" and more controversial and less widely used, would apply in severe cases of impulse control disorders or command hallucinations.

In such cases, the expert witness would have to establish a diagnosis of a recognized mental (e.g. schizophrenia) or neurological (e.g. brain tumour) disorder and then show that this disorder led to a functional impairment resulting in "cognitive insanity". If neuroimaging comes into play such as in the case of a brain tumour, the challenge is not so much to prove the abnormality of the imaging finding (which is generally defined by widely accepted clinical criteria), but its contribution to the act in question (for example, using the counterfactual thought experiment whether someone, given their primary character, would have been likely to commit such an act before the disease in question developed). A more difficult situation from the neuroimaging perspective arises when a defendant does not present with a qualitative and clinically recognized abnormality but an expert would like to argue that he or she was incapable of moral reasoning because of a quantitative abnormality, for example reduced perfusion in prefrontal areas that are crucial for this type of reasoning. This type of reasoning, which has been applied in a number of criminal cases (Werner et al., 2019), faces major methodological challenges (including determining normative values, and accounting for the plastic nature of the human brain) (Jones et al., 2013). Although it has so far had limited use in actual legal practice this move to more quantitative measures that might be able to place people on a spectrum of criminal responsibility, rather than using the classical dichotomy of sanity vs. insanity, is interesting from a theoretical perspective. It can be seen as part of a move to deconstruct the multifactorial causation of delinquent behaviour that incorporates both biological and psychosocial vulnerabilities. The argument could go like this—why should a perpetrator whose act was "caused" by a brain tumour receive more leniency from a court that the perpetrator whose act was "caused" by a concatenation of psychosocial adversity, early drug abuse, resulting brain damage and lack of access to support services? Ultimately, a causal chain of physical events exists for any human action. Whether or not this basic fact of physical causation of human action is relevant for their praise- or blameworthiness is a matter of intense philosophical debate (Greene & Cohen, 2004), but whatever one's position in this debate, it should not matter what kind of brain process caused a criminal offence. Either everyone is capacity responsible for their actions whatever their brain scan reveals because physical causation is irrelevant for the question of responsibility, or nobody is (this is the position advocated by Greene and Cohen). The latter case does not remove the rationale for punishment (only for its retributivist aspect), nor does it remove the need to differentiate between offenders on the basis of their brain state (see next sections), but it moves these considerations into the domain of liability responsibility.

Liability and Responsibility

In Vincent's taxonomy, "liability (responsibility)" refers to the way in which an offender will be treated by society, what liability will be imposed on them. In the criminal context, this is mainly about the justification for and determination of the right level of punishment. Of the five purposes of punishment identified by Vincent, all but the first (retribution) can still very much apply even in a model that denies everyone capacity responsibility. These are "general and specific deterrence", "reform and education of the offender", "quarantine of dangerous people to protect society", and "expression of society's solidarity with victims by publicly condemning offenders' actions" (Vincent, 2010). Deterrence, reform and education and protection of society from dangerous individuals often require close liaison between legal practitioners and clinical experts from psychiatry, psychology and neuroscience. Contributions of neuroscience to considerations of deterrence and education and reform will be discussed in this section, whereas the rationale for the quarantining of dangerous individuals will be discussed in the subsequent section on prevention of re-offending.

Let us consider the justification of punishment by its deterrent properties through a few examples. The patient who committed an offence because a frontal brain tumour interfered with the normal functioning of areas that are essential for moral reasoning and/or impulse control, does not need to be punished (at least not under the auspices of deterrence) after a successful operation of the brain tumour and restitution of his or her previous level of functioning. Specific deterrence does not apply because the patient's risk of re-offending is not conceivably higher than that of the general population, and general deterrence does not apply either, because it is reasonable to assume that it will not be possible to deter someone (a hypothetical other patient) with this kind of brain tumour with the prospect of a jail term. The patient simply will not have the ability to judge the outcomes of his or her action, and whether there is a chance that he or she might go to jail for it does not enter into his or her considerations. Conversely, the person who committed a similar offence because of the chain of adverse upbringing and circumstances outlined in the previous section might be punished by a jail sentence because it is reasonable to assume that his or her brain (and that of others in a similar situation) is capable of processing information about reward and punishment and make this a guiding principle of the person's decision-making. Finally, a perpetrator who, like the patient with the brain tumour, is not able to take potential future punishment into consideration when planning and performing certain actions, but who does not have the prospect of a definitive cure, for example a patient suffering from severe learning disability, would not be punished on grounds of deterrence, but may still need to be committed to an institution for the protection of the public.

The question of educational prospects and reform of the perpetrator is closely coupled to that of deterrence. Most people for whom deterrence does not work because of severe learning disability or another severe and enduring mental and/or neurological disorder will also show limited if any responsiveness to attempts at education and reform. Because they were driven in their actions by simple stimulus–response mechanisms uncontrolled by higher-order planning mechanisms the main educational strategy will generally be one of behaviour analysis and modification, rather than one that employs more complex cognitive strategies. The scientific understanding of the brain and behaviour, both in general and in relation to the individual perpetrator, can help formulate the appropriate strategy for education and reform although the advice from educational and behavioural psychologists will generally be more relevant than that from neuroscientists. The situation changes if there is a treatable underlying condition that would transform the perpetrator's understanding of and ability to follow moral norms and legal rules. For example, in the aforementioned case of the perpetrator with the frontal brain tumour, removal of the brain tumour will be the key step to reform, so much so that very little if any additional education is needed in order to reintegrate him or her into normal societal processes. Similar, for a perpetrator with a chronic delusional disorder, successful psychiatric treatment of this disorder may be the key to reformability. In such cases, the advice of a clinical neuroscientist or psychiatrist may be sought to determine the best disposal and determine the risk of re-offending.

Prevention of Re-offending

This brings us to the last major section of the legal process to which neuroscience could make a meaningful contribution, prognosis, and prevention. The assessment of re-offending risk is generally a complex and multifactorial procedure, but in some cases, such as that of the perpetrator with the brain tumour mentioned above who committed an illegal act that was completely out of character, a close causality between a treatable disease and the act can be determined, resulting in a positive prognostic assessment after successful treatment. At the other end of the spectrum of treatability, it is also conceivable that the expert would conclude, for example in a case of an incurable brain tumour or a progressive neurodegenerative disease, that a patient remains permanently dangerous until he or she is physically or cognitively so impaired as to become incapable of any independent action. Mental disorders sit between these two extremes because they are generally neither completely curable nor incurable, and thus, any remaining risk owed to a treated but not completely cured illness would be matter of degree. Beyond categorical diagnoses one can also consider neuroscience-based quantitative parameters such as local brain volume or metabolism as potential predictors of recidivism (Delfin et al., 2019). However, the required longitudinal studies are very difficult to conduct, and the necessary independent validation of predictive models in new cohorts poses considerable logistic and ethical challenges.

In public and legal debate the issue of preventive detention, that is, the confinement of a person beyond the term of their sentence for public protection, often arises in the context of sexual offending but also in other cases of violent crimes. Sexual offenders pose a considerable risk of re-offending, current treatments are not particularly effective in reducing this risk, and risk predictions in individual cases have a large margin of error (Dennis et al., 2012). The question for legislators is then whether sexual offenders should be released from prison when they have served their term and be a risk to the community or whether they should be detained as long as they pose a risk—and thus potentially indefinitely? Over the last 30 years legislatures and law courts, for example in many States of the USA, several Australian provinces, New Zealand, Germany, Scotland and England have increasingly favoured the second option (Janus, 2013, McSherry & Keyzer, 2009). In the USA, all State laws providing for confinement of sexually violent predators follow the criteria set out by the Supreme Court in the case *Kansas v. Hendricks* (1997). In order to be legally detained in a special facility beyond the term of the original conviction the offender needs to have.

"(1) a history of sexually harmful behavior; (2) a mental abnormality that produces an impairment of control over sexually harmful behavior; (3) a prediction of future sexually dangerous behavior" (Janus, 2013).

Clinical neuroscience as such rarely comes into the determination of these criteria because the "abnormality" concerned is generally not of an identifiable neurological nature (and it if it, the question of treatability becomes paramount). However, criteria 2 and 3 are within the purview of the forensic psychiatrist or psychologist. Some sexually violent predators fulfil diagnostic criteria for an identified mental disorder from the group of paraphilias such as paedophilia or sexual sadism disorder (Linden, 2019). However, the criteria are not confined to these patients with a classical psychiatric diagnosis because they use the broader (and controversial) term of "mental abnormality"(McSherry & Keyzer, 2009) which may be a universal category for those assessed as posing a risk of future sexually dangerous behaviour (and it is not automatically the case that those with a diagnosis are more dangerous than those without). In many cases, the application of these criteria by courts (and the legislation they have to follow) thus becomes more of a politico-legal than a purely medical-scientific matter. In addition to the contribution to diagnosis and risk assessment, neuroscience and psychiatry might also have a role in the future development of treatments that might reduce re-offending rate. A classical debate concerns the use of brain surgery for sexual offenders (Linden, 2014), which is becoming a topical issue again because of the advances of deep brain stimulation (Fuss et al., 2015) but still far away from practical implementation. However, if patients are detained preventively on the basis of a medical model that assigns them an abnormality that leads to them posing a risk to the community they should also have access to any treatment that might cure or mitigate their condition (Merkel, 2007). Thus, it would be desirable to have a debate about new treatments enabled by recent advances in clinical and basic neuroscience that could potentially reduce the risk of sexual violence or other types of violent offending.

Summary and Conclusions

Whereas the role of neuroimaging and other neurotechnologies for determining the *actus reus* is currently very limited, neuroscience (in its broad definition that includes mental health) has considerable relevance for the evaluation of the *mens rea* and insanity as well as for questions of disposal and prognosis. Most of the questions that are currently posed to a forensic psychiatrist will still be relevant if legal systems abandon the classical intuitions of blameworthiness and retribution and move to a purely consequentialist system of punishment (Greene & Cohen, 2004). After all, questions of capacity responsibility and liability (sensu Vincent) are closely entwined, and although retribution would disappear from the scope of the latter in a consequentialist system, the other aspects of punishment would remain. Practical differences would probably be limited: In a consequentialist system, most offenders found to have diminished responsibility in the current system would be punished under the auspices of education and reform and protection of the public, rather than deterrence. As argued above, proper assessment of these categories frequently involves the evaluation of neural and psychological criteria for reformability and re-offending, which will be relevant to any society whatever its philosophical views on moral responsibility. Individual assessments of capacity for moral reasoning and impulse control as well as the investigation of the underlying neural mechanisms will most probably still play a major role in both retributivist and consequentialist legal systems.

References

Delfin, C., Krona, H., Andiné, P., Ryding, E., Wallinius, M., & Hofvander, B. (2019). Prediction of recidivism in a long-term follow-up of forensic psychiatric patients: Incremental effects of neuroimaging data. *PLoS ONE, 14*(5), e0217127. https://doi.org/10.1371/journal.pone.0217127.

Dennis, J. A., Khan, O., Ferriter, M., Huband, N., Powney, M. J., & Duggan, C. (2012). Psychological interventions for adults who have sexually offended or are at risk of offending. *Cochrane Database of Systematic Reviews, 12,* CD007507. https://doi.org/10.1002/14651858.CD007507.pub2.

Farah, M. J., Hutchinson, J. B., Phelps, E. A., & Wagner, A. D. (2014). Functional MRI-based lie detection: Scientific and societal challenges. *Nature Reviews Neuroscience, 15*(2), 123–131. https://doi.org/10.1038/nrn3665.

Franssen, S., Jansen, A., van den Hurk, J., Roebroeck, A., & Roefs, A. (2020). Power of mind: Attentional focus rather than palatability dominates neural responding to visual food stimuli in females with overweight. *Appetite, 148,* 104609. https://doi.org/10.1016/j.appet.2020.104609.

Fuss, J., Auer, M. K., Biedermann, S. V., Briken, P., & Hacke, W. (2015). Deep brain stimulation to reduce sexual drive. *Journal of Psychiatry and Neuroscience, 40*(6), 429–431.

Greene, J., & Cohen, J. (2004). For the law, neuroscience changes nothing and everything. *Philosophical Transactions of the Royal Society of London. Series B, Biological Sciences, 359*(1451), 1775–1785. https://doi.org/10.1098/rstb.2004.1546.

Janus, E. S. (2013). Preventive detention of sex offenders: The American experience versus international human rights norms. *Behavioral Sciences & the Law, 31*(3), 328–343. https://doi.org/10.1002/bsl.2059.

Jones, O. D., Wagner, A. D., Faigman, D. L., & Raichle, M. E. (2013). Neuroscientists in court. *Nature Reviews Neuroscience, 14*(10), 730–736. https://doi.org/10.1038/nrn3585.

Linden, D. (2012). Overcoming self-report. Possibilities and limitations of brain imaging in psychiatry. In S. Richmond, G. Rees & S. J. L. Edwards (Eds.), *I know what you're thinking: Brain imaging and mental privacy*. Oxford: Oxford University Press.

Linden, D. (2014). *Brain control*. Basingstoke: Palgrave.

Linden, D. (2019). *The biology of psychological disorders* (2nd ed.). Red Globe Press: [distributor] Springer-Verlag Berlin and Heidelberg GmbH & Co. KG.

McSherry, B., & Keyzer, P. (2009). *Sex offenders and preventive detention: Politics, policy and practice*. Annandale, NSW: Federation Press.

Merkel, R. (2007). *Intervening in the brain: Changing psyche and society*. Berlin: Springer.

Poldrack, R. A. (2006). Can cognitive processes be inferred from neuroimaging data? *Trends in Cognitive Sciences, 10*(2), 59–63. S1364-6613(05)00336-0. https://doi.org/10.1016/j.tics.2005.12.004.

Rusconi, E., & Mitchener-Nissen, T. (2013). Prospects of functional magnetic resonance imaging as lie detector. *Frontiers in Human Neuroscience, 7*, 594. https://doi.org/10.3389/fnhum.2013.00594.

Saarimäki, H., Gotsopoulos, A., Jääskeläinen, I. P., Lampinen, J., Vuilleumier, P., Hari, R., Sams, M., & Nummenmaa, L. (2016). Discrete neural signatures of basic emotions. *Cerebral Cortex, 26*(6), 2563–2573. https://doi.org/10.1093/cercor/bhv086.

Simon, R. J., & Ahn-Redding, H. (2006). *The insanity defense the world over*. Lanham, MD and Oxford: Lexington Books.

Uncapher, M. R., Boyd-Meredith, J. T., Chow, T. E., Rissman, J., & Wagner, A. D. (2015). Goal-directed modulation of neural memory patterns: Implications for fMRI-based memory detection. *Journal of Neuroscience, 35*(22), 8531–8545. https://doi.org/10.1523/JNEUROSCI.5145-14.2015.

Vilares, I., Wesley, M. J., Ahn, W. Y., Bonnie, R. J., Hoffman, M., Jones, O. D., et al. (2017). Predicting the knowledge-recklessness distinction in the human brain. *Proceedings of the National Academy of Sciences of the United States of America, 114*(12), 3222–3227. https://doi.org/10.1073/pnas.1619385114.

Vincent, N. A. (2010). On the relevance of neuroscience to criminal responsibility. *Criminal Law and Philosophy, 4*, 77–98.

Werner, R. A., Savoie, B., Javadi, M. S., Pomper, M. G., Higuchi, T., Lapa, C., & Rowe, S. P. (2019). From the reading room to the courtroom-the use of molecular radionuclide imaging in criminal trials. *Journal of the American College of Radiology, 16*(11), 1612–1617. https://doi.org/10.1016/j.jacr.2019.05.001.

David Linden is full Professor of Translational Neuroscience and Scientific Director of the School for Mental Health and Neuroscience. He is a Psychiatrist at Maastricht University Medical Centre. His specialist clinical areas include neuropsychiatry, genetic syndromes in psychiatry, mood disorders, psychosis, and alcohol dependence. His research focuses on mechanisms and treatment of mental and neurodegenerative disorders. His group combines neuroimaging, cognitive neuroscience, genetics, and clinical research in order to develop new biological models and find new treatment targets.

Neuroscience and Dangerousness Evaluations: The Effect of Neuroscience Evidence on Judges. Findings from a Focus Group Study

Georgia Gkotsi

Introduction

Neuroscientific research on the relationship between neurobiology and antisocial behaviour has rapidly grown over the last two decades, causing vivid discussions on potential uses of neuropredictive models of violence as indicators of future dangerous behaviour. Forensic neuroprediction, i.e. uses of recent developments in neuroscience in criminal justice contexts, in order to improve predictions about an individual's risk of (re-)engaging in antisocial conduct, is one of the most intriguing challenges for our legal system. While neuroscience holds the promise of adding predictive value to existing risk assessment tools, its potential use for justice purposes raises a variety of scientific, epistemological, legal and ethical issues. One of the relevant concerns is related to the

G. Gkotsi (✉)
Faculty of Law, National and Capodistrian University of Athens, Athens, Greece
e-mail: ggkotsi@law.uoa.gr

S. Ligthart et al. (eds.), *Neurolaw*, Palgrave Studies in Law, Neuroscience, and Human Behavior,
https://doi.org/10.1007/978-3-030-69277-3_2

prejudicial nature of neuroscientific data. The latter could be unduly persuasive despite the lack of scientific support for their use to diagnose cognitive or behavioural impairments. Several studies in experimental psychology have demonstrated a number of cognitive effects arising from exposure to neuroimaging data which may bias judgements and lead to (mis)interpretations that can affect legal decisions.

If interpreted by judges or juries as evidence of the fact that the defendant is constitutively, irremediably dangerous, neuroscientific evidence in criminal settings could eventually open the door to unjustifiably severe punishment and/or an overly aggressive use of preventive detention for potentially dangerous individuals in the name of public safety.

A key question is how judges and juries are going to perceive and evaluate this kind of data, and if they are going to give too great a weight to them. The prejudicial impact of neuroscientific evidence remains an open empirical question to be examined. Several studies have been conducted on the potentially biasing effects of neuroscientific evidence, especially in the context of mock juries. These studies gave mixed results concerning the over-evaluation of neuroscience evidence and their potentially biasing effects (Brown & Murphy, 2009; McCabe & Castel, 2008; Weisberg et al., 2008; but see Schweitzer et al., 2011, as cited by Greely & Farahany, 2019). Up to now, little research has been conducted on the impact of neuroscientific evidence and its effects on the decisions made by trial judges (Moulin et al., 2018; Cheung & Heine, 2015; Fuss et al., 2015). In order to investigate this issue, we conducted a pilot study with focus groups, whose preliminary results are presented in this article.

Neuroscience as a Better Tool for Evaluations of Future Dangerousness/recidivism Risk[1]?

Up to now, four generations of tools have been created to evaluate an offender's risk to society: First-generation clinical evaluations, structured or non-structured, have been accused of lack of solid methodology, objectivity and false results, either false positives or false negatives. Violence risk assessment based on clinical evaluation does not have a particularly good track record, and some experts have even suggested that relying on psychiatric predictions of violence is tantamount to "flipping coins in the courtroom" (Nadelhoffer & Sinnott-Armstrong, 2012).

Second-generation tools, based on actuarial prediction, depend on statistical analysis of a subject's objective information. Actuarial methods rely on specific variables that are weighted in predetermined ways (statistical methods of evaluation include Violence Risk Appraisal Guide VRAG, Static-99 and Static-2002). Actuarial methods have better predictive value but are criticized for being based on unchangeable factors that hinder any prospective of change and evolution of the subject (Quinsey et al., 2006).

Third-generation tools, the so-called Professional structured judgement tools (such as HCR-20, LSI-R, The Hare Psychopathy Checklist-Revised-PCL-R) assemble estimates of risk by reviewing and scoring a set list of empirically validated risk factors known to be associated with violence. In this approach, structure is imposed on which risk factors should be considered and how they should be measured. The weighing of their importance into an assigned level of risk is considered as the result of clinical judgement (Philips, 2012).

Recently, a new generation of risk assessment tools has emerged. Fourth-generation tools place emphasis on the individual's strengths,

---

[1] Dangerousness is a rather mysterious and paradoxical notion, since it implies at once the affirmation of a quality immanent to the subject, and a mere probability, a quantum of uncertainty, given that the proof of the danger can only be provided after the fact, should the threatened action actually occur (Castel, 1991). Characterized mainly by vagueness and polysemy, the notion of dangerousness is abandoned in many countries, especially in Anglosaxon countries since 1980, in favour of the notion of the risk (Pratt, 2001), while in continental countries the term "dangerousness" is still used.

We use the terms "dangerousness" and "recidivism risk" interchangeably.

taking into consideration protective factors that reduce the chances of manifestation of (sexually) violent behaviour or recidivism. "SAPROF" (Structured Assessment of Protective Factors for Violence Risk), designed in 2007, is the most widely used fourth-generation tool and its results are combined with results derived from third-generation risk assessment tools. Fourth-generation tools not only include risk-need assessment but also integrate the assessment with a case management plan. Treatment is adjusted to the individual's needs, while assessments of risk factors (including protective factors) are periodically adjusted and reevaluated (Abbiati et al., 2017; Bonta & Wormith, 2008).

Despite the fact that latest-generation tools have evolved, taking into consideration protective risk factors and incorporating techniques for intervention and treatment strategies, methods of predictions of future dangerousness/recidivism risk in general remain controversial and are constantly criticized for lack of accuracy and reliability (Calcedo-Barba, 2006; Friend, 2003; Douglas et al., 2017; Fazel et al., 2012). In this context, neuroscience emerges as a novel and scientific way to help psychiatrists make a step towards developing more exact tools for the neuroprediction of violent behaviour. Recent findings regarding structural and/or functional brain damage correlated with the manifestation of violent behaviour have paved the way for the use of neuroscience knowledge and techniques in forensic settings and raise increasing interest for forensic psychiatrists, neuroscientists (Simpson, 2012; Silva, 2007; Delfin et al., 2019; Poldrack et al., 2018) and legal scholars (Redding, 2006; Morse, 2015). Several authors consider recent neuroscientific discoveries as a useful tool able to provide justice with credible evidence that will improve accuracy and reduce errors in psychiatric expertise (Aggarwal, 2009; Witzel, 2012; Silva, 2006; Simpson, 2012; Nadelhoffer & Sinnott-Armstrong, 2012).

In hope of improving the accuracy of existing tools, several studies are taking place, seeking to explore potential uses of recent brain imaging developments for taking a step towards the possibility of developing new tools for the "neuroprediction" of violence for forensic uses.

In 2013, the first prospective forensic neuroprediction study was published by Aharoni et al., in which researchers used fMRI in a group of 96 male prisoners and followed them at prison release for 4 years. The study results indicated that those individuals showing low activity in a brain region associated with decision-making and action (the Anterior Cingulate Cortex) are more likely to being rearrested within 4 years of release. According to the study, the risk of recidivism is more than double in individuals showing low activity in that region of the brain than in individuals with high activity in that region. The results of this study, according to the authors, suggest a "potential neurocognitive biomarker for persistent antisocial behavior" (Aharoni et al., 2013).

Recent studies by Kiehl et al. (2018) suggested that "models that combined psychological, behavioral, and neuroimaging measures provided the most robust prediction of recidivism".

Kiehl and his team set out to discover whether brain age—an index of the volume and density of grey matter in the brain—could help predict re-arrest. The results verify the utility of brain measures in predicting future behaviour and suggest that reduced grey matter in the anterior temporal lobes, amygdala, and orbital frontal cortex was more helpful in predicting rearrest than was chronological age.

Delphin et al. (2019) conducted a long-term (ten-year average time at risk) follow-up study to include neuroimaging data in the prediction of recidivism in a forensic psychiatric sample.

Researchers studied whether the inclusion of resting-state regional cerebral blood flow measurements leads to an incremental increase in predictive performance over traditional risk factors. A Baseline model with eight empirically established risk factors, and an extended model which also included resting-state regional cerebral blood flow measurements from eight brain regions were compared using several predictive performance metrics.

These aforementioned studies suggest that brain scans can theoretically help determine whether certain convicted persons are at an increased risk of reoffending if released. However, given the increasing concern for public safety, there is discussion and several concerns are raised about uses of neuroscientific data for the assessment of levels of risk posed by offenders (Gaudet et al., 2016; Gkotsi & Gasser, 2016; Morse, 2015; Petersen, 2014).

Ethical and Legal Concerns: Neuroscience as a Risk for Offenders' Individual Rights

The use of neuroscientific data in criminal cases raises a variety of scientific, epistemological, legal and ethical issues. One of the relevant issues concerns the extent to which neuroscientific data, used in the context of forensic psychiatry, can influence the judgement and the outcome of decisions made by judges and jurors.

The scholarly empirical literature on the effects of such evidence is mixed (Shen et al., 2017). Recent research suggests that although neuroscience information may be persuasive under certain conditions (Scurich & Appelbaum as cited by Shen et al., 2017), brain images themselves are not independently persuasive. As a result, as Shen et al. comment: "research going forward is likely not to address 'Does neuroscientific evidence affect outcomes?' (inviting a binary Yes/No answer), but rather 'How much and under what circumstances does neuroscientific evidence affect outcomes?'".

Within the current social crimino-political situation, judges, confronted with the pressure to ensure public security, could consider neuroscience as a reliable tool, indispensable in assessing an offender's dangerousness. Within this context, fears are expressed that judges might rely too heavily on neuroscientific evidence and opt for heavier sentences or perpetuating post-sentence measures, on the basis of the offenders' neurobiological profile that allegedly proves that the latter are predisposed to criminal behaviour and thus more likely to recidivate.

According to some authors, if neuroscientific data are interpreted as evidence of dangerousness, it is highly likely that the judge will impose heavier sentences and/or—in European continental systems—security or therapeutic measures, which can be indeterminate in length. Thus, the use of neuroscience in criminal psychiatric expertise might be risky for defendants. This is the "double-edged sword" effect of neuroscience in court, outlined by several commentators (Barth, 2007; Farahany & Coleman, 2009): even if research and neuroscientific data are introduced by defence lawyers in criminal proceedings through a psychiatric expertise with the aim to prove diminished responsibility, these same data can be interpreted by judges as an indication of dangerousness of the defendant and lead to long-term sanctions/measures based on the assumption of a high probability of recidivism in subjects with brain dysfunction. Although a recent empirical study's findings controvert the image of neuroscience evidence as a double-edged sword (Denno, 2015), discussion continues on the subject.

The use of neurobiology as a neurocognitive biomarker also risks labelling offenders on the basis of their neurobiological profile and discriminating against them in everyday life after release (Fuchs, 2006; Bedard, 2017).

This tendency could be exacerbated by the fact that neuroscientific evidence is often perceived as more objective, reliable, and "scientific" evidence, despite the limitations and difficulties of reliably connecting current brain function to future behavioural patterns. Images—and neuroimages in particular—can have a more profound effect on jury and judge determinations than verbal testimony, as several studies of social psychology have shown (Gurley & Marcus, 2008; McCabe & Castel, 2008; Weisberg et al., 2008; Kulynych, 1997), although Shen et al. (2017) found that "neuroscientific evidence does affect outcomes, but it has a weaker effect than the strength of the case".

Thus, courts might be compelled to use neuroscience to ground responsibility and dangerousness assessments, which could open the door to a more aggressive use of preventive detention for potentially dangerous individuals, undermining the principle of proportionality that lies at the core of criminal sentencing, that is, the idea that the punishment of a certain crime should be in proportion to the severity of the crime itself.

The Effect of Neurobiological Evidence on Judges and Jurors: Findings from Studies

Recent psychological studies indicate that neuroscientific knowledge or neuroimages moderately increase the perceived scientific credibility of accompanying information (Weisberg et al., 2008; Kulynych, 1997), and that "lay readers infer more scientific value for articles including brain images than those that do not, despite their lack of sufficient scientific evidence and regardless of whether the article included reasoning errors or not" (McCabe & Castel, 2008).

According to several studies, brain images in particular are likely to impact evaluations of an argument's credibility (Gurley & Marcus, 2008; McCabe & Castel, 2008). This is linked to the so-called seeing is believing effect, which has been demonstrated by recent research in experimental psychology and suggests the existence of cognitive bias concerning the reliability and validity of a scientific study, when it is accompanied by a photograph or image (McCabe & Castel, 2008). Even though these results are not confirmed by meta-analyses (Schweitzer et al., 2011; Michael et al., 2013), these studies raise questions about the possibility of neuroscientific data being perceived by jurors and judges as more "scientific" than other types of evidence.

As demonstrated by Moulin et al. (2018), including neuroscience evidence in an expert report may impact the way the report is assessed by non-specialists, such as judges. The study showed that the presence of neuroscience data in an expert report affects judges' perceptions of the quality, credibility and scientificity of the report, and the persuasiveness of the evidence is provided.

Although in some cases neuroscience data actually does have some evidential value and correctly affects perceptions, the question is if this kind of evidence is sometimes unduly persuasive. The overly persuasiveness of neuroscientific evidence has been attributed to the fact that the collection of this kind of data requires a complex technological process, which apparently attributes to the findings greater scientific value. A different explanation could be related to the tendency of non-experts/laypeople to consider sciences, such as psychiatry and social sciences as less reliable, less valid and less rigorous than "hard" sciences such as physics and biology (Munro & Munro, 2014; Simonton, 2009), or their tendency to prefer simple, reductionist, explanations for complex phenomena (Crommelinck, 1995).[2]

---

[2]Others suggest that neuroscience evidence is more likely to have a prejudicial effect when structural neuroimaging techniques are used as evidence in court: structural abnormalities are more likely to influence judgements and mitigate punishment decisions than functional abnormalities, as the latter have less causal potency than the structural ones. See Choe, S. Y.

In any case, even though more recent studies suggest that neuroscience is not as biasing as feared (Roskies et al., 2013; Michael et al., 2013; Farah & Hook, 2013; Schweitzer et al., 2011, 2013) the prejudicial impact of neuroscientific evidence, i.e. its capacity to often unduly affect perceptions of judges' remains an open empirical question to be examined (Nadelhoffer et al., 2012; Gruber & Dickinson, 2012).

The Effect of Neuroscientific Evidence on Judges: First Findings of a Pilot Study from Focus Groups

Aiming to explore this issue, i.e. the way in which neuroscientific evidence is perceived and the extent to which it can be prejudicial, we conducted a study with focus groups consisting of judges, lawyers, neurologists and psychiatrists, whose aim was to detect eventual "biases" as to the persuasiveness, objectivity and scientific quality of experts' opinions that include neuroscientific tools and findings, especially in comparison with traditional, clinical psychiatric expert evaluations.

This focus group study was conducted in the context of a larger research project[3] whose aim was to examine uses of neuroscientific evidence in criminal trials through psychiatric expert opinions, and more specifically to examine the way in which neuroscientific evidence is perceived by judges, as well as its impact on the kind and length of the sentence imposed on mentally or/and neurologically impaired offenders. The research addressed, among others, the following issues: what is the Judges' opinion on psychiatric expert opinions in general and on expert opinions that incorporate neuroscientific knowledge in particular? What is the judges' perception of the notion of "dangerousness" in general and if/how they associate it with mental illness and neurobiological deficits?

(2014). Misdiagnosing the impact of neuroimages in the courtroom. *UCLA Law Review, 61*, 1502–1548.

[3]The research was funded as a project by the Greek State Scholarships Foundation (IKY): Gkotsi Georgia, "Criminal treatment of mentally ill offenders in the age of neuroscience: uses of neuroscientific data in psychiatric expert opinions" 2016–2018.

The larger methodology of the research included analysis of relevant case law and a combination of qualitative research methods, such as focus groups and interviews with judges.

Selecting the Focus Groups' Method

Focus groups are a qualitative research technique designed to explore a range of perceptions and views of a research subject through the participants' own perspective (Morgan, 1996b; Krueger & Casey, 2010). A focus group is a gathering of deliberately selected people who participate in a planned discussion. It explicitly uses group interaction as part of the method and allows members to interact and influence each other during the discussion. During a focus group, a group discussion is held where participants discuss a specific topic, exchanging views and commenting on their experiences (Kitzinger, 1995), thus, the method is particularly useful for exploring people's knowledge and experiences.

For the purpose of this study and given that neurolaw is an unexplored field in the Greek legal context, focus groups were considered as a suitable method for bringing together all professionals involved in criminal trials (judges, defence lawyers, experts—neurologists and psychiatrists), in order to elicit their perceptions on uses of neurobiological data in criminal trials, in the context of a psychiatric expertise and to detect potential bias on behalf of the judges concerning the use of neurobiological data. Used as a pilot study, focus groups featured the participants' thoughts, opinions and knowledge on the subject, through interaction, highlighting the participants' point of view on the researched subject and reflecting their role in a criminal trial.

Focus groups can generally be used as a method either autonomously, or in combination with qualitative or quantitative methods (Morgan, 1996a). In the context of our larger research, this method was used as a pilot study, in combination with the qualitative method of individual interviews with judges (Lambert & Loiselle, 2008; Merton, 1987). By bringing out the reasoning, way of thinking and diversity of judges' concerns, focus groups' results were taken as sources of new ideas and contributed to identifying appropriate themes and formulating questions for interviews with judges, which constituted the next stage of our research.

Team Design and Composition

Two focus group sessions were organized and took place in December 2017–March 2018 in Athens, Greece with nine selected individuals. Participants were divided into two teams by professional occupation: legal scholars (lawyers and judges) and psychiatrists/neurologists. Since each discipline uses very different methodology and jargon, the division of the participants in two homogeneous groups according to their professional occupation was considered necessary in order to ensure participants' comfort in sharing their thoughts and knowledge in a familiar group, and to achieve the efficient performance of the group dynamics (Krueger & Casey, 2010; Ritchie et al., 2013, p. 190). In addition, division in two teams was considered necessary in order to prevent a possible infiltration of the analysis by prejudices or stereotypes that one professional group would have against the other, and which would eventually result in an alteration of the data.

The group of lawyers included two Judges servicing in the Athens Court of First Instance, a defence lawyer specializing in the defence of mentally ill persons, a lawyer specializing in bioethics and a lawyer and social anthropologist providing legal aid and advocacy for people suffering from mental health problems. The second group was composed of two neurologists and two psychiatrists with experience as experts in Courts. The discussions were coordinated by the researcher. Participants were briefed on the purpose and subject of the study and completed a consent form by which they agreed to the recording of the discussion and their anonymity was ensured. A plan of semi-guided general questions (discussion guide) provided the basis and stimulus for the discussion.[4] Thematic analysis was chosen as the method of analysis that

[4]Judges were asked to generally comment on the increasing tendency to introduce behavioural genetics and neuroimaging techniques in attempts to exculpate criminal defendants and to mitigate defendants' culpability and punishment. Questions/issues for discussion also included

systematically attempts to identify, analyze and report patterns within data and thereby provide cognitive access to collective significations and experiences (Braun & Clarke, 2006).

For the purposes of this chapter, we will present some findings from the discussion that took place in the "legal" group, consisting of judges and lawyers.

Findings from the Focus Group Consisting of Judges and Lawyers

Four main issues emerged during the discussion: these issues concerned (i) the extent to which participants think that neuroscientific data can contribute to improving the quality of psychiatric expert opinions (section "Contribution of Neuroscientific Data to the Improvement of the Quality of Psychiatric Expert Opinions"), (ii) The principle of free evaluation of evidence and its power when an expert opinion incorporating neuroscientific data is introduced in a criminal trial (section "The Relationship Between an Expert Opinion Incorporating Neuroscientific Data as Means of Evidence and the Principle of Free Evaluation of Evidence"), (iii) The issue of dangerousness and how participants correlate dangerousness with mental illness and neurobiological data (section "The Issue of Dangerousness: Correlation Between Dangerousness, Mental Illness and Neurobiological Data") and (iv) The use of neurobiological data as evidence of reduced responsibility in the context of a defence strategy (section "Neurobiological Data as Evidence of Reduced Responsibility in the Context of a Defence Strategy")

recent trial cases in the context of which neuroimaging techniques were used as evidence in a criminal court, as well as studies which explored uses of recent developments in neuroscience in order to improve predictions about an individual's risk of (re-)engaging in antisocial conduct.

Contribution of Neuroscientific Data[5] to the Improvement of the Quality of Psychiatric Expert Opinions

Concerning the degree to which neuroscience could improve the quality of psychiatric testimony, participants were divided into two subgroups: on the one hand, judges seemed convinced that neuroscientific data could potentially serve as a valuable tool for improving the quality and reliability of psychiatric expert opinions and contribute to a safer diagnosis of a mental illness and to a more exact evaluation of the defendant's clinical status in general (section "Judges: Improving the Reliability of Psychiatric Expert Opinions with Neuroscientific Data"). On the other hand, other participants were more sceptical about the use of this data in criminal proceedings, pointing out several some scientific, legal and conceptual limitations related to their use in criminal settings (section "Scepticism About Improving the Reliability of Psychiatric Expert Opinion Using Neuroscientific Tools—"Pseudo-Objectification". Scientific Limitations and Epistemological Difficulties").

Judges: Improving the Reliability of Psychiatric Expert Opinions with Neuroscientific Data[6]

According to judges who participated in the focus group, neuroscientific data can improve the quality of psychiatric opinion in two ways: contributing to a better diagnosis of the psychiatric mental illness and to

---

[5]We employ the term "neuroscientific data" as a generic term including general information derived from published neurobiological studies, related to the relationship between brain and behaviour, as well as data obtained from brain imaging techniques. These techniques can be either structural (magnetic resonance imaging (MRI), computerized axial tomography (CAT)), or functional, such as electroencephalogram (EEG), functional magnetic resonance imaging (fMRI), Positron Emission Tomography (PET) and Single-Photon Emission Computed Tomography (SPECT).

[6]In an inquisitorial criminal justice system, procedural guarantees serve a different conceptual logic than adversarial systems, i.e. a conceptual priority has to be given to requirements concerning the 'quality' of the non-partisan state official expert (Decaigny, 2014). Experts must have previously acquired knowledge and skills that allow them to fulfil their mission and to be appointed by judges. In Greece, a country of inquisitorial system, experts are registered in official lists of experts, are commissioned by investigating judges and prosecutors and cannot be commissioned by the defence or the civil parties.

a more exact evaluation of the defendant's clinical condition in general and assisting in the formulation of the judges' opinion.

Safer Mental Illness Diagnosis—More Exact Evaluation of the defendant's Clinical Condition with Neuroscientific Data

The discussion revealed a general mistrust on the part of the judges towards clinical psychiatric expert opinions. As the main reasons of their mistrust, judges mention the ambiguity and lack of scientific objectivity in the documentation of expert opinions and the gaps often encountered in the diagnosis of psychiatric illnesses. They also criticize forensic psychiatry generally for lack of a well-defined methodology, which often leads to erroneous—either false negatives or false positives—results with regard to dangerousness evaluations.

Neurological and biological data, as opposed to "traditional" psychiatric data, are considered by judges participating in the focus group to be of better quality and more reliable, while clinical psychiatric examination is considered inaccurate and not particularly reliable. The foundation of a mental illness on an organic, cerebral basis with the aid of neuroimaging tests, lends credence to psychiatric assessment and is therefore considered by judges as more "objective" and scientifically valid.

As one of the judges noticed:

> ... there are gaps in traditional psychiatric methods regarding the diagnosis and, with the rise of neuroscience, these gaps become evident... so I think that this tendency to use neuroscientific tools should be considered positively, because it gives a more adequate portrait of the examined person. Neuroscientific techniques ... would help as a safe method of diagnosis .... (E.E. Judge)

Assistance in the Formulation of the Judges' Opinion

Special reference is made by judges to the difficulties of decision-making whenever specialized knowledge (from a different discipline—psychiatry and neurology in the case) is required. The situation in which they find themselves when having to comprehend and evaluate a psychiatric opinion is emphatically described as "floating in an ocean". Judges highlight the great responsibility they are charged with when judging the future of an accused person. In this context, data derived from neuroimaging techniques and examinations are perceived as an extremely useful tool that should be integrated in psychiatric opinions, in combination with other tests and methods, in order to assist and provide security to the judge and ultimately contribute to more effective administration of justice and provide legal certainty.

In a very characteristic quote, M.B., judge, comments:

> ... the use of neuroscientific techniques would prevent us from floating in the ocean of a psychiatric expertise .... combining neuroscience with clinical examination could offer a lifeline in this ocean. Anything that objectifies this vague expertise makes you feel more secure about the administration of justice. (M.B. Judge)

And they add:

> I personally feel that it will untie my hands, it will help me understand this person's problem. (M.B. Judge)

The terms "safety board" and "will untie my hands" emphasize the judge's feeling of helplessness, whenever they are required to base their decision on specialized knowledge with which they are not familiar. An expert opinion incorporating neuroscientific data is perceived by judges as "objective", based on undisputed technical scientific evidence. As one of the judges points out, not only can this kind of knowledge not be ignored, but it is part of a judge's duty to consider latest technology data, in their quest to find the truth.

M.B, Judge, comments:

> ... science is evolving, we cannot ignore it, I would not have a clear conscience if I ignored it completely ... this kind of knowledge can help the judge establish legal certainty.

Scepticism About Improving the Reliability of Psychiatric Expert Opinion Using Neuroscientific Tools—"Pseudo-Objectification". Scientific Limitations and Epistemological Difficulties

The rest of the group's participants, i.e. lawyers, seemed more sceptical on the reliability of this kind of data and aware of the scientific limitations of neuroimaging technologies.

Lawyers point to the early state of development of neuroimaging technology, as well as the lack of neurobiological diagnostic markers. They question the relevance of group derived data for one person and they specifically refer to the difficulties in establishing causal links in attributing a type of behaviour to a specific brain structure or dysfunction. During the discussion, it was also mentioned that genetic polymorphisms, such as the MAOA gene[7] and, more generally, information regarding genetic predispositions cannot provide precise answers for specific individuals in a personalized way.

In addition, lawyers made extensive reference to the epistemological limitations and difficulties of communication between the judge and the psychiatrist-expert and highlighted the need to distinguish scientific reasoning from legal reasoning.

As one of the lawyers characteristically comments:

> It is one thing how a judge is called upon to judge and how a scientist, a doctor or a biologist, reasons. The judge must judge in black white at the end. Scientists never reason like that. (B.T., Lawyer)

[7]According to part of the scientific literature, MAOA-uVNTR polymorphism points to a "genetic vulnerability" thought to predispose the subject to exhibiting aggressiveness when challenged or excluded socially, see Caspi, A., McClay, J., Moffitt, T. E., Mill, J., Martin, J., Craig, I. W., & Taylor, A. (2002). Poulton R. Role of genotype in the cycle of violence in maltreated children. *Science, 297*(5582), 851–854.

The Relationship Between an Expert Opinion Incorporating Neuroscientific Data as Means of Evidence and the Principle of Free Evaluation of Evidence

Articles 177 and 178 of the Greek Code of Criminal Procedure establish the principles of the free evaluation of evidence and the principle of the free use of any evidentiary means. Together, they constitute the principle of moral proof, according to which judges interpret facts, including scientific facts, "in light of their reasoned intimate conviction" (Byk, 2012). Under this principle, judges are free to formulate their opinion without being bound by legal rules of evidence. Thus, according to the prevailing view in theory, expert opinions are freely assessed by the court and experts' conclusions are not and should not be binding for the judge. If they were binding, the expert, whose role is to assist the judge, would substitute the latter, jeopardizing the constitutional requirement for justice administration by the courts (Paraskevopoulos & Kosmatos, 2013). As in any inquisitorial system, in the Greek legal system, scientific data can help to construct the "legal truth", which, however, may not be reduced to these facts and judges are free to distance themselves from scientific data. As a result, according to the dominant view in theory, judges are free not to take into consideration the outcome of an expertise, as long as they provide justification for this decision (Konstantinides, 2009). However, with regard to the justification requirement, it has often been commented that the judge, lacking the necessary specialized knowledge, cannot, de facto, put forward scientific arguments in order to contradict or to reject the expert's findings. This is the reason why it has been partly supported in theory that an expert's opinion as an evidentiary means should be binding (Kaiafa-Gbandi, 1983; Androulakis, 1973).

This issue emerged during the discussion concerning the use of neuroscientific data in criminal settings, given that knowledge that comes from neuroscientific methods and techniques is particularly technical and specialized knowledge.

Again, two subgroups were formulated within the group, expressing two opposite opinions on this matter: on the one hand, lawyers who participated in the focus group express concerns that whenever neuroscientific data is incorporated in an expert opinion, rarely will judges be in a position to freely assess it, as it will be extremely hard for judges to refute this kind of knowledge (section "Free Assessment of an Expert Opinion is not Possible When It Integrates Neuroscientific Data"). On the other hand, judges do not consider this kind of evidence as a threat to their service, and they emphasize the primacy of legal reasoning and their ability to resist to the "seductive" effect of (neuro)scientific evidence (section "Neuroscientific Tools Are Not Likely to Unduly Affect Judges' Reasoning").

Free Assessment of an Expert Opinion is not Possible When It Integrates Neuroscientific Data

The question was raised by lawyers participating in the focus group, according to which, the use of neuroscientific data in criminal proceedings through psychiatric expert opinion may constitute a "*trap*" for judges: the term "trap" is employed in order to indicate that there is a strong possibility that judges accept this type of data undisputedly. This undisputed acceptance may lie in the interpretation of neuroscientific data by judges as objective technical-scientific, scientifically valid and reliable data. But, according to lawyers, even if judges appear unconvinced and uncertain as to the neuroscientific expert's opinion's credibility, they may eventually end up by accepting it, as—lacking the necessary specialized knowledge—they will not be in a position to refute it.

Concern is also being expressed about judges' opinions being substituted by scientific data, which is identified with an automated way of administering criminal justice.

L.A., lawyer, expresses this concern as follows:

> It is like giving a sort of tool to the judge, which inactivates their judgment and decides at their place if the accused person will be responsible or not. This seems both unscientific and outrageous to me. You're rendering the judge obsolete.

Concerns are also expressed about the possibility that this type of data is used to the detriment of defendant's rights, resulting in a reversal of the burden of proof in violation of the presumption of innocence and the right to a fair trial.

In the end, the lawyers participating in the group conclude that the support that such data can provide to the judge is not substantial, but only psychological in nature. According to them, neuroscientific tools can act as an "authority" that helps judges only psychologically, as it lifts the burden of a difficult decision, but in reality it hampers judicial work, undermining free evaluation of evidence and acting as a substitute of legal judgement. Lawyers basically express concern that the legal reasoning is going to be replaced by scientific reasoning.

As V.T., lawyer, comments:

> I think that the more scientific tools you put in the game of evidence, the more you drive the judge away from making that decision. After all, you take away their responsibility. (V.T., Lawyer)

Neuroscientific Tools Are Not Likely to Unduly Affect Judges' Reasoning

Judges participating in the focus group take a defensive stance to this issue, seeking to establish the primacy of legal reasoning and emphasize the independence and autonomy of their service: they reply that there is no danger of replacement of their judgement by neuroscience, as it is through legal reasoning that they will be able to refute an expertise/psychiatric testimony. To lawyers' concerns about the risk of judges being overly persuaded and basing their decision on elements of questionable credibility, they oppose the legal framework and established case law, which, according to them, provides them with a means of defence against questionable expert opinions.

E.E., Judge, defends the power of legal reasoning and its ability to resist questionable science in the courtroom in the following words:

> According to the Constitution, it is us that must make the decision and it is us who are called upon to reason...What an expert will say will help me, but the expert will not make the decision for me .... And the

> argumentation/confrontation will be based on a legal criterion, not on a scientific one. (E.E. Judge)

The Issue of Dangerousness: Correlation Between Dangerousness, Mental Illness and Neurobiological Data

Neurobiological Data as Evidence of Dangerousness

In general, dangerousness and recidivism risk of the offender are explicitly mentioned by the two judges participating in the focus group as an important criterion that plays a crucial role in deciding which type of sentence or custodial or therapeutic measure to choose. Taking into account the public opinion in their decision, judges are especially concerned about their duty to protect society.

This concern is emphatically expressed in M.B., Judge's comment:

> Judges have a mission, that is, to protect public safety…dangerousness, as a factor, does exist in the mind of a judge and is always taken into account, in fact, it is the main factor which is taken into account. (M.B., Judge)

During the discussion, judges strongly correlated dangerousness with mental illness and with schizophrenia in particular, an approach which accords with the social stereotypes of the "violent mentally ill offender" that associate severe mental illness—and especially schizophrenia with violence.

E.E., Judge, characteristically mentions:

> Mental illness, to a certain extent, carries a very high degree of risk. You can't release a schizophrenic person. This person objectively constitutes a danger to society …. (E.E., Judge)

Judges interpret the existence of neurobiological abnormalities as indicative of a different biological structure between "violent" and "non-violent" individuals. Brain damage is considered by judges as permanent damage which results in the loss of ability to control impulse.

E..E, Judges, mentions:

> People suffering from a degenerative nervous system have been observed to have impulses and urges, more than normal people, that's for sure.

Potentially dangerous individuals appear to be grouped/characterized as biologically different on the basis of their dysfunctional brain. And it is this particular characteristic, the dysfunctional or damaged offender's brain, that justifies an individualized sentence and is crucial to the judges' decision to impose either a custodial or therapeutic measure. Judges tend to believe that neuroscience can help differentiate between a dangerous person and a mentally ill person.

E.E. one of the Judges' comment indicates that neuroscientific data could indirectly be perceived by judges as indicative of an offender's dangerousness and affect their decision accordingly.

> If a person is indeed completely incompetent because of their damaged brain, this will personally help me understand this person's problem and judge accordingly whether this person should be in custody in case they're dangerous, or receive treatment if they suffer from a disease. (E.E., Judge)

Neurobiological Data as Evidence of Reduced Responsibility in the Context of a Defence Strategy

From the lawyers' point of view, neuroscientific data could prove useful in the context of a defence strategy, in cases where the court is sceptical on the existence of an alleged mental disorder and considers an existing expert opinion to be unreliable. In this respect, these elements could serve as a tangible, "organic" proof of the existence of a mental disorder that excludes (or reduces) responsibility.

Thus, although having previously acknowledged the fact that neurobiological data cannot objectify an existing psychiatric disease, lawyers are aware of the appeal of this kind of data to judges and do not hesitate to give it a try in the context of a defence strategy.

However, lawyers are aware of the limitations of using this kind of data in criminal settings and express concerns about whether it will benefit the defendant. They are very much aware of the fact that this kind of data is open to interpretation, which makes it flexible and allows its use in the criminal process as strategic tool both for defence and prosecution.

As one of the lawyers, D.S., comments:

> I would use this kind of tool if the court did not believe that my client truly suffers from a mental illness and had doubts or considered the expert opinion unreliable... However, either as a defense lawyer, or as the plaintiff's lawyer, if one was to use it against me, I would definitely have a lot to say to the court about its unreliability.

Indeed, as shown by case studies (Gkotsi et al., 2019) conflicting expert testimony and radically different interpretations of the same neuroscientific data suggest that the latter are open to interpretation by neuroscientists and are susceptible to being presented and interpreted by experts according to the legal side they represent.

This is related to the "double edged sword effect" (Barth, 2007), according to which neuroscience could indeed lead to defendants being found less blameworthy, but such evidence could also backfire, if judges conclude that the neuroscience shows the defendant is constitutively, irremediably dangerous, and hence must be locked away for a longer period of time to protect the public.

Discussion

As a result of the discussion and interaction that was developed in the "legal" group two "sub-groups" were created, judges and lawyers, who disagreed and confronted each other on several of the discussed issues. This confrontation between lawyers and judges reflects their distinguished roles in the criminal proceeding. Lawyers' primary concern focuses on protecting the interests of their clients and they frequently express concern that neuroscientific data may be used to the detriment of the latter. Judges respond defensively to the concerns expressed by lawyers that judges may misinterpret or be "seduced" by such evidence and support the use of this kind of evidence in criminal settings, believing it will substantially help them in formulating their opinion.

While lawyers are sceptical as to uses of neuroscientific evidence in court, judges are unaware of the scientific limitations of neuroimaging techniques and consider neuroscience to be a valuable tool, one that will "untie their hands" as one of them mentions in a very characteristic way. When they have to make a decision in a field where specialized knowledge is required, judges consider data obtained from neuroimaging techniques to be highly reliable and scientifically valid, as opposed to data obtained through "traditional" clinical psychiatric examination. Judges have high expectations from neuroscience, hoping that it will contribute to the "objectification" of a seemingly opaque discipline, such as psychiatry. Neurobiological data, due to their supposed biological basis, are considered as able to objectify psychological and psychiatric data and thus as "physical" support for psychological and psychiatric conclusions. The foundation of a mental illness on an organic, cerebral basis lends credence to psychiatric assessment, which is therefore considered as more "objective" and scientifically valid, when enriched with findings from neuroimaging techniques or information about the brain. Therefore, this kind of data is hoped to make psychological and psychiatric evaluations more reliable, more coherent and more scientific.

References of judges to neuroscience being able to prevent them *"from floating in the ocean of a psychiatric expertise .... and objectifying vague (traditional psychiatric) expertises* indicate that to the judges' mind, a valid medical approach should be embedded in the positivist tradition, according to which, valid knowledge is identified with scientific knowledge. The latter must be cleared from any metaphysical element derived from "traditional" psychiatry, which, to the judges' mind, constitutes a cloudy scientific landscape.

Even though they acknowledge their potential contribution to a defence strategy, lawyers participating in the focus group are not enthusiastic about the use of neurobiological data in criminal settings, acknowledging their scientific limitations and the fact that this kind of data is open to interpretation, which makes them eligible to serve as strategic tools both for defence and prosecution.

The discussion also shows that non-specialists tend to categorize neurologically impaired individuals by their dysfunctional brain, as having a biologically different structure. According to them, the ability to examine the perpetrators' brain reveals useful information that allows for an individualizing sentencing and facilitates the decision to impose a custodial or therapeutic measure.

There is a common expectation of all the participants in the legal team (judges and lawyers) that neuroscientific data will suggest new ways of treatment and prove useful in selecting the most appropriate therapeutic treatment/measure. This could indirectly point to the fact that participants associate dangerousness with a brain disease or dysfunction which can be treated. Participants believe that new knowledge about the brain could lead to an increased adoption of individualized, socio-rehabilitative measures, which will contribute to reducing recidivism of offenders upon release to the community. In this context, dangerousness could be considered as a clinical condition with a neurological basis that can be identified and treated.

This approach brings in mind current discussions about the uses of neuroscience for assessing the possibility of treatment of perpetrators. It is associated with current discussions on the uses of neuroscience for evaluating a perpetrator's "treatability" and raises the issue of a return to the therapeutic approach to crime promoted through neuroscience, as revived by numerous recent studies on the neurobiological basis of violent behaviour and crime (Raine, 2013).

Throughout the entire discussion, tension is evident between Science and Law which are perceived by participants as polarized disciplines, antagonizing each other. The two disciplines are in constant competition, which reflects their particular relationship and their different social functions and purposes. Law pursues the abstract idea of justice, whereas science attempts to describe and, ultimately, explain real phenomena. Yet, at a lower level, law does deal with real circumstances and events, and so cannot avoid recourse to evidence, including scientific evidence (Eastman & Cambell, 2006). The discussion stresses the need to delineate the scope of each discipline, but also the possibilities of cooperation. What is therefore needed is to overcome the communication barriers between the judge and the psychiatrist-expert and effectuate a "translation" of the results of neuroscientific research and techniques presented to a court, to the legal language.

Limitations

In the present study, the focus groups' method was selected as a pilot study, with the aim to make a preliminary investigation of the perceptions of professionals involved in criminal proceedings concerning the use of neurobiological data in criminal contexts and to investigate the potential "bias" as to persuasiveness, objectivity and scientific quality of experts' opinions that include neuroscientific tools and findings. The focus group study is part of a larger research and its findings must be combined with findings from individual interviews with judges which constituted the next stage of the research.

The organization of a single session does not allow for the generalization of results. In order to confirm and enrich the findings of this pilot focus group, it is necessary in the future to organize more group sessions per professional category and possibly a final session involving a joint group of legal scholars and neuroscientists—psychiatrists that will interact.

As far as the composition of the teams is concerned, it would be useful that the team of legal scholars be enriched, apart from lawyers, with judicial officers of all levels (Presidents of the Court of First Instance, judges in Courts of Appeals, Prosecutors) of all ages and experience, in order to examine the extent to which experience, age and qualification influences the perceptions of the judicial officers regarding the credibility, scientificity and objectivity of neurobiological data used as evidence.

Conclusion

The preliminary findings from a first focus group suggest that judges do tend to consider neuroscientific data as credible, objective and scientific, useful pieces of evidence that will assist them in deciding. At the same time, this kind of evidence is, to their mind, able to give an exact insight to an offender's clinical and neurological condition and thus guide them in imposing a suitable sentence or measure. On the other hand, though acknowledging its potential use as a defence strategy tool, lawyers are more sceptical concerning their use in criminal trials, taking into account the interpretative nature of this kind of evidence.

In addition, judges interpret the existence of neurobiological abnormalities as indicative of a different biological structure between "violent" and "non-violent" individuals, which suggests that neuroscientific data introduced in a criminal setting may be interpreted as strong evidence of dangerousness, based on the high probability of recidivism of brain-injured offenders.

Judges distinguish and put special emphasis on neuroscientific data as a decisive and objective factor on which dangerousness assessments could reliably be based upon, disregarding other factors which should combinedly be taken into account in assessing a person's future risk of committing new crimes. However, reducing dangerousness to a single factor, to a specific neurobiological structure in the case, can lead to stigmatization of people with brain malfunctions, who could be defined as dangerous, based simply on a trait they possess: their defected brain. In this context, despite the fact that neuroscience findings can assist, to an extent, in assessing an offender's future dangerousness, there is a danger of returning to a simplistic explanation of violent behaviour, if neuroscientific evidence is presented by the experts or understood by judges as the ultimate scientific and objective tool, able to prove a causal link between some structural or functional brain abnormality and the propensity to manifest criminal behaviour.

The alleged ability to detect dangerousness-based exclusively on brain malfunctions maximizes social expectations of identifying a category of potentially dangerous individuals and exercising social control on them.

Hence, members of the legal profession must be trained in how to recognize the strengths and weaknesses of this new type of evidence expert report. Only by correctly assessing neuroscience data, while remaining aware of its potential impact on their evaluative and decision-making processes, will they be able to exploit its potential contribution to evaluating future behaviours (Moulin et al., 2018).

It is undisputed and suggested by all the participants—judges themselves included—that there is a need to train judges on this matter. Judges, and legal professionals in general, must be trained in how to evaluate this new type of expert report without allowing its perceived objectivity to influence their critical faculties. As judges may overestimate the importance of neurobiological deficits for the assessment of responsibility or the prediction of criminal behaviour, their training would aim to inform them on the limitations and the interpretative nature of this kind of knowledge and make them more vigilant as to the interpretation and meaning attributed to neuroscientific data. Only by remaining aware of neuroscientific data's potential impact on their decision-making processes, will judges be able to exploit their potential contribution to evaluating and explaining behaviours (Moulin et al., 2018).

Finally, assessing the role of the neuroscience for the evaluation of responsibility and dangerousness of a mentally ill person, we should bear in mind that the issue of distinguishing "normal" from mentally ill people is not an exclusively epistemic matter, but to a certain extent, normative. The bipole "normal - pathological" (Canguilhem, 1966) is a fundamental form of organization of medical knowledge that organizes corresponding forms of intervention on the phenomena of health and disease, however, from the standpoint of the philosophy of science, the definition of the concept of "normal" remains fluid and polysemous, directly related to the gnosiotheoretical system in which it emerges and used each time. Whether neuroscientific findings will help solve this issue, offering data which will be useful to distinguish pathological from non-pathological people on the basis of the brain remains uncertain, as it will most likely continue to be, to a large extent, a theoretical/philosophical and normative discussion.

In the current socio-political context, where expectations vis-à-vis psychiatrists are particularly high, often based on the hope of anticipating and eliminating all kinds of risk, we should be aware of the risk of distorting the meaning of neuroscientific data with unrealistic and arbitrary interpretations, resulting in the imposition of heavier sentences and preventive detention for some categories of criminal offenders, based on their defective brain.

Acknowledgements The author would like to thank judges and doctors who participated in this research.

References

Abbiati, M., Azzola, A., Palix, J., Gasser, J., & Moulin, V. (2017). Validity and predictive accuracy of the structured assessment of protective factors for violence risk in criminal forensic evaluations: A Swiss cross-validation retrospective study. *Criminal Justice and Behavior, 44*(4), 493–510.

Aggarwal, N. K. (2009). Neuroimaging, culture, and forensic psychiatry. *Journal of the American Academy of Psychiatry and the Law Online, 37*(2), 239–244.

Aharoni, E., Vincent, G. M., Harenski, C. L., Calhoun, V. D., Sinnott-Armstrong, W., Gazzaniga, M. S., & Kiehl, K. A. (2013). Neuroprediction of future rearrest. *Proceedings of the National Academy of Sciences, 110*(15), 6223–6228.

Androulakis N. (1973). *The expert—Psychiatrist in criminal trial*. Poinika Chronika ΚΓ΄, 327.

Barth, A. S. (2007). Double-edged sword: The role of neuroimaging in federal capital sentencing. *American Journal of Law & Medicine, 33,* 501–522.

Bedard, H. L. (2017). The Potential for bioprediction in criminal law. *Science and Technology Law Review, 18*.

Bonta, J., & Wormith, S. J. (2008). Risk and need assessment. In G. McIvor & P. Raynor (Eds.), *Developments in social work with offenders* (pp. 131–152). London, UK: Jessica Kingsley Publishers.

Braun, V., & Clarke, V. (2006). Using thematic analysis in psychology. *Qualitative Research in Psychology, 3*(2), 77–101.

Brown, T., & Murphy, E. (2009). Through a scanner darkly: Functional neuroimaging as evidence of a criminal defendant's past mental states. *Stan. L. Rev., 62,* 1119.

Byk, C. (2012). Justice et expertise scientifique: Un dialogue organisé dont il faut renouveler les fondements. In O. Oullier (Ed.), *Le cerveau et la loi: analyse de l'émergence du neurodroit* (pp. 151–167). Paris: Département Questions sociales, Centre d'analyse stratégique.

Calcedo-Barba, A. (2006). The ethical implications of forensic psychiatry practice. *World Psychiatry, 5*(2), 93–94.

Canguilhem, G. (1966). *Le normal et le pathologique*. PUF.

Castel, R. (1991). From dangerousness to risk. In G. Burchell, C. Gordon, & P. Miller (Eds.), *The foucault effect: Studies in governmentality* (pp. 281–298). Chicago: The University of Chicago Press.

Cheung, B. Y., & Heine, S. J. (2015). The double-edged sword of genetic accounts of criminality: Causal attributions from genetic ascriptions affect legal decision making. *Personality and Social Psychology Bulletin, 41*(12), 1723–1738.

Crommelinck, M. (1995). Quand la trace des souvenirs se dévoile au fond d'une coupelle. À propos du réductionnisme et des neurosciences. *Revue philosophique de Louvain, 93*(1), 140–175.

Decaigny, T. (2014). Inquisitorial and adversarial expert examinations in the case law of the European court of human rights. *New Journal of European Criminal Law, 5*(2), 149–166.

Delfin, C., Krona, H., Andiné, P., Ryding, E., Wallinius, M., & Hofvander, B. (2019). Prediction of recidivism in a long-term follow-up of forensic psychiatric patients: Incremental effects of neuroimaging data. *PloS ONE, 14*(5).

Denno, D. W. (2015). The myth of the double-edged sword: An empirical study of neuroscience evidence in criminal cases. *Boston College Law Review, 56*, 493.

Douglas, T., Pugh, J., Singh, I., Savulescu, J., & Fazel, S. (2017). Risk assessment tools in criminal justice and forensic psychiatry: The need for better data. *European Psychiatry, 42*, 134–137.

Eastman, N., & Campbell, C. (2006). Neuroscience and legal determination of criminal responsibility. *Nature Reviews Neuroscience, 7*(4), 311–318.

Farah, M. J., & Hook, C. J. (2013). The seductive allure of "seductive allure." *Perspectives on Psychological Science, 8*(1), 88–90.

Farahany, N. A., & Coleman, J. E., Jr. (2009). Genetics, neuroscience, and criminal responsibility. In N. A. Farahany (Ed.), *The impact of behavioral sciences on criminal law* (pp. 183–240). New York: Oxford University Press.

Fazel, S., Singh, J. P., Doll, H., & Grann, M. (2012). Use of risk assessment instruments to predict violence and antisocial behaviour in 73 samples involving 24 827 people: Systematic review and meta-analysis. *BMJ, 345*, e4692.

Friend, A. (2003). Keeping criticism at bay: Suggestions for forensic psychiatry experts. *Journal of the American Academy of Psychiatry and the Law Online, 31*(4), 406–412.

Fuchs, T. (2006). Ethical issues in neuroscience. *Current Opinion in Psychiatry 19*(6), 600–607.

Fuss, J., Dressing, H., & Briken, P. (2015). Neurogenetic evidence in the courtroom: A randomised controlled trial with German judges. *Journal of Medical Genetics, 52*(11), 730–737.

Gaudet, L. M., Kerkmans, J. P., Anderson, N. E., & Kiehl, K. A. (2016). Can neuroscience help predict future antisocial behavior. *Fordham Law Review, 85,* 503.

Gkotsi, G. M., & Gasser, J. (2016). Neuroscience in forensic psychiatry: From responsibility to dangerousness. Ethical and legal implications of using neuroscience for dangerousness assessments. *International Journal of Law and Psychiatry, 46,* 58–67.

Gkotsi, G. M., Gasser, J., & Moulin, V. (2019). Neuroimaging in criminal trials and the role of psychiatrists expert witnesses: A case study. *International Journal of Law and Psychiatry, 65,* 101359.

Greely, H. T., & Farahany, N. A. (2019). *Neuroscience and the criminal justice system.* Annual Review of Criminology.

Gruber, D., & Dickerson, J. A. (2012). Persuasive images in popular science: Testing judgments of scientific reasoning and credibility. *Public Understanding of Science, 21*(8), 938–948.

Gurley, J. R., & Marcus, D. K. (2008). The effects of neuroimaging and brain injury on insanity defenses. *Behavioral Sciences & the Law, 26*(1), 85–97.

Kaiafa-Gbandi, M. (1983). Should findings from expert opinions be binding for criminal courts? *Armenopoulos 1983: 1046* (in Greek).

Kiehl, K. A., Anderson, N. E., Aharoni, E., Maurer, J. M., Harenski, K. A., Rao, V., … & Kosson, D. (2018). Age of gray matters: Neuroprediction of recidivism. *NeuroImage: Clinical, 19,* 813–823.

Kitzinger, J. (1995). Qualitative Research: Introducing Focus Groups. *BMJ, 311*(7000), 299–302.

Konstantinides A. (2009). Addiction and expert opinion. Probative value of the expert opinion. In *Criminality and Rule of Law* (Nomiki Vivliothiki, Ed.).

Krueger, R. A., & Casey, M. A. (2010). Focus group interviewing. In J. S. Wholey, H. P. Hatry, & K. E. Newcomer (Eds.), *Handbook of practical program evaluation* (3rd ed.). San Francisco, CA: Jossey-Bass.

Kulynych, J. (1997). Psychiatric neuroimaging evidence: A high-tech crystal ball? *Stanford Law Review,* 1249–1270.

Lambert, S. D., & Loiselle, C. G. (2008). Combining individual interviews and focus groups to enhance data richness. *Journal of Advanced Nursing, 62*(2), 228–237.

McCabe, D. P., & Castel, A. D. (2008). Seeing is believing: The effect of brain images on judgments of scientific reasoning. *Cognition, 107*(1), 343–352.

Merton, R. K. (1987). The focussed interview and focus groups: Continuities and discontinuities. *The Public Opinion Quarterly, 51*(4), 550–566.

Michael, R. B., Newman, E. J., Vuorre, M., Cumming, G., & Garry, M. (2013). On the (non) persuasive power of a brain image. *Psychonomic Bulletin & Review, 20*(4), 720–725.

Morgan, D. L. (1996a). *Focus groups as qualitative research* (Vol. 16). Sage.

Morgan, D. L. (1996b). Focus groups. *Annual Review of Sociology, 22*(1), 129–152.

Morse, S. (2015). Neuroprediction: New technology, old problems. In *Bioethica Forum* (Vol. 8, p. 128).

Moulin, V., Mouchet, C., Pillonel, T., Gkotsi, G. M., Baertschi, B., Gasser, J., & Testé, B. (2018). Judges' perceptions of expert reports: The effect of neuroscience evidence. *International Journal of Law and Psychiatry, 61,* 22–29.

Munro, G. D., & Munro, C. A. (2014). "Soft" versus "hard" psychological science: Biased evaluations of scientific evidence that threatens or supports a strongly held political identity. *Basic and Applied Social Psychology, 36*(6), 533–543.

Nadelhoffer, T., & Sinnott-Armstrong, W. (2012). Neurolaw and neuroprediction: Potential promises and perils. *Philosophy Compass, 7*(9), 631–642.

Nadelhoffer, T., Bibas, S., Grafton, S., Kiehl, K. A., Mansfield, A., Sinnott-Armstrong, W., & Gazzaniga, M. (2012). Neuroprediction, violence, and the law: Setting the stage. *Neuroethics, 5*(1), 67–99.

Paraskevopoulos, N., & Kosmatos, K., (2013). *Drugs: Interpretation by article of the criminal and procedural provisions of the "Addictive Substances Act"*, ed. Sakkoulas (3rd ed.) (in Greek).

Petersen, T. S. (2014). neuropredictions. *The Journal of Ethics, 18,* 137–151.

Philips, R. (2012). Predicting the risk of future dangerousness. *Virtual Mentor, 14*(6), 472–476.

Poldrack, R. A., Monahan, J., Imrey, P. B., Reyna, V., Raichle, M. E., Faigman, D., & Buckholtz, J. W. (2018). Predicting violent behavior: What can neuroscience add? *Trends in Cognitive Sciences, 22*(2), 111–123.

Pratt, J. (2001). Dangerosité, risque et technologies du pouvoir. *Criminologie,* 101–121.

Quinsey, V. L., Harris, G. T., Rice, M. E., & Cormier, C. A. (2006). Criticisms of actuarial risk assessment. In V. L. Quinsey, G. T. Harris, M. E. Rice, & C. A. Cormier (Eds.), *The law and public policy. Violent offenders: Appraising and managing risk* (pp. 197–223). American Psychological Association.

Raine, A. (2013). *The psychopathology of crime: Criminal behavior as a clinical disorder*. Elsevier.

Redding, R. E. (2006). The brain-disordered defendant: Neuroscience and legal insanity in the twenty-first century. *American University Law Review, 56,* 51.

Ritchie, J., Lewis, J., Nicholls, C. M., & Ormston, R. (Eds.). (2013). *Qualitative research practice: A guide for social science students and researchers.* Sage.

Roskies, A. L., Schweitzer, N. J., & Saks, M. J. (2013). Neuroimages in court: Less biasing than feared. *Trends in Cognitive Sciences, 17*(3), 99–101.

Schweitzer, N. J., Baker, D. A., & Risko, E. F. (2013). Fooled by the brain: Re-examining the influence of neuroimages. *Cognition, 129*(3), 501–511.

Schweitzer, N. J., Saks, M. J., Murphy, E. R., Roskies, A. L., Sinnott-Armstrong, W., & Gaudet, L. M. (2011). Neuroimages as evidence in a mens rea defense: No impact. *Psychology, Public Policy, and Law, 17*(3), 357.

Shen, F. X., Twedell, E., Opperman, C., Krieg, J. D. S., Brandt-Fontaine, M., Preston, J., ... & Carlson, M. (2017). The limited effect of electroencephalography memory recognition evidence on assessments of defendant credibility. *Journal of Law and the Biosciences.*

Silva, J. A. (2006). The relevance of neuroscience to forensic psychiatry. *Journal of the American Academy of Psychiatry and the Law Online, 35*(1), 6–9.

Silva, J. A. (2007). The relevance of neuroscience to forensic psychiatry. *Journal—American Academy of Psychiatry and the Law, 35*(1), 6.

Simonton, D. K. (2009). Varieties of (scientific) creativity: A hierarchical model of domain-specific disposition, development, and achievement. *Perspectives on Psychological Science, 4*(5), 441–452.

Simpson, J. R. (Ed.). (2012). *Neuroimaging in forensic psychiatry: From the clinic to the courtroom.* Chichester, West Sussex, UK: Wiley-Blackwell.

Weisberg, D. S., Keil, F. C., Goodstein, J., Rawson, E., & Gray, J. R. (2008). The seductive allure of neuroscience explanations. *Journal of Cognitive Neuroscience, 20*(3), 470–477.

Witzel, J. (2012). Implications of neuroimaging for dangerousness assessment. In J. R. Simpson (Ed.), *Neuroimaging in forensic psychiatry: From the clinic to the courtroom* (pp. 195–200). Chichester, West Sussex, UK: Wiley-Blackwell.

Dr. Georgia Gkotsi is a Research Fellow at the Faculty of Law of the University of Athens, Greece. After receiving a law degree from the University of Athens, she completed a Master's in Philosophy of Law and Bioethics at the same University, followed by a Master's in Comparative Law at the Universite Paris 1 - Pantheon Sorbonne. She received her PhD from the University of Lausanne, Switzerland. Her dissertation dealt with the ethical and legal implications of the use of neuroimaging techniques in criminal courts. Her research expertise lies in the area of mental health law, human rights of mentally disabled persons, neurolaw, and bioethics.

[illegible]

The Need for a Partial Defence of Diminished Capacity and the Potential Role of the Cognitive Sciences in Helping Frame That Defence

Paul Catley

Introduction

Criminal laws are implicitly based on a premise that people could act otherwise and are to some degree responsible for their actions. Unless one adopts a hard determinist stance, one will be likely to view people, at least to some limited extent, as exercising choice in their actions. It is therefore justifiable to hold them both responsible and blameworthy, for their actions/choices. The exceptions of duress, automatism and insanity apply in rare circumstances, but generally the criminal law proceeds on the basis that people are presumed responsible for their actions. There may then be questions as to whether they have the necessary *mens rea* for the offence charged. However, assuming both *actus reus* and *mens rea* are proved, guilt will be established, and the next step will be to progress to sentencing.

P. Catley (✉)
Open University Law School, The Open University, Milton Keynes, UK
e-mail: paul.catley@open.ac.uk

S. Ligthart et al. (eds.), *Neurolaw*, Palgrave Studies in Law, Neuroscience, and Human Behavior,
https://doi.org/10.1007/978-3-030-69277-3_3

Running parallel to this theoretical model of the criminal law's operation, we also know that many who are in prison have mental health problems and/or brain abnormalities/injuries (Allely, 2016; Bronson & Berzofsky, 2017; Friestad & Kjelsberg, 2009; Schilz et al., 2013; Shiroma et al., 2012). While some of these may have developed in prison, many will have existed at the time that they offended. These individuals may not have met the criminal law's narrow insanity criteria when they offended, but their behaviour may well have been impacted by their mental health and/or cognitive functioning, for example through reduced reasoning powers and/or impulse control. This chapter will argue that the criminal law in all jurisdictions should recognize an intermediate level of criminal responsibility between those deemed not criminally responsible and those held to be criminally responsible. Such an intermediate level exists in some jurisdictions, for example the Netherlands (see Meynen, 2016, pp. 145–148[1]), but such jurisdictions are rare.

Responsibility, Culpability and Fair Fault Ascription

In *Placing Blame* Michael Moore states:

> My own theory is that criminal law is a functional kind [of judgement] whose function is to attain retributive justice. Retributive justice demands that those who deserve punishment get it. To deserve punishment, two things are necessary: one must have done a wrongful action, and one must have done so culpably. (1997, p. 33)

Moore explains that in his view culpability relates only to those who are "morally culpable" (1997, p. 35) and links this to a requirement that to be held responsible one "must be sufficiently rational and autonomous to be a moral agent" (1997, p. 403). I agree with Moore that to deserve

[1] Since Meynen's chapter the Netherlands has moved from five levels of criminal responsibility to three.

punishment a person must have committed a wrongful act and done so culpably. In this chapter, I will focus on the idea of culpability and in doing so will interpret culpability as more than the simple *mens rea* element of an offence, such as intention, recklessness or negligence, and instead incorporate within culpability an additional and separate requirement of blameworthiness.

This element of blameworthiness stems in part from Herbert Hart's idea that criminal liability can be built on:

> the simple idea that unless a man has the capacity and a fair opportunity or chance to adjust his behaviour to the law its penalties ought not to be applied to him. (1968, p. 181)

My contention is that rather than a simple binary divide between individuals with "capacity and a fair opportunity or chance to adjust" their behaviour and those who do not; there should be an intermediate category for those who have limited capacity or limited opportunity or chance to adjust their behaviour. These individuals are still blameworthy but are less blameworthy than those who can reasonably easily adjust their behaviour.

Blameworthiness and Hard Determinism

Joshua Greene and Jonathan Cohen suggest that "neuroscience will probably have a transformative effect on the law (…) by transforming people's moral intuitions about free will and responsibility" (2004, p. 1775). Their argument is that "Free will, as we ordinarily understand it, is an illusion" (2004, p. 1783). They accept that determinism precludes the idea of an "uncaused causer" with the effect that:

> the problem of attributive free will arises. To see something as a responsible moral agent, one must first see it as having a mind. But, intuitively, a mind is, among other things, an uncaused causer. Consequently, when something is seen as a mere physical entity operating in accordance with deterministic physical laws, it ceases to be seen, intuitively, as a mind.

> Consequently, it is seen as an object unworthy of moral praise or blame. (2004, p. 1782)

Neuroscience, they argue, "will undermine people's common sense, libertarian conception of free will and the retributivist thinking that depends on it" (2004, p. 1776). If Greene and Cohen are correct, then arguing for diminished capacity is absurd. If nobody is blameworthy, then gradations of blameworthiness are illusory.

It is sometimes claimed that the importance for the law of the libertarian concept of free will has been addressed and dismissed. Such claims will often cite Stephen Morse's (2007) article *The Non-Problem of Free Will in Forensic Psychiatry and Psychology.* While it is true that Morse argues that free will is not a problem for the law, he acknowledges that "incompatibilist hard determinism (…) would obliterate the possibility of responsibility altogether" (2007, p. 204). As he explains:

> If determinism is true, the people we are and the actions we perform have been caused by a chain of causation over which we mostly had no rational control and for which we could not possibly be responsible.[2,3] (2007, p. 213)

Free will may or may not be an illusion. But, if it is an illusion, it is one, for the time being at least, in which society believes. Though Morse tries to argue that free will is not important for the law, he acknowledges that any form of responsibility, including criminal responsibility, rests

[2]Arguably Morse's use of the words "mostly" and "rational" in the above quotation are misleading (and should be omitted) as they imply some degree of control which if hard determinism is true would be erroneous.

[3]Whilst Morse here accepts that "if determinism is true" we cannot be responsible, elsewhere in his writing he adopts an apparently different stance. For example, in a later work he claims: "free will is not a legal criterion that is part of any doctrine and it is not even foundational for criminal competence or responsibility. Criminal law doctrines are fully consistent with the truth of determinism or universal causation that allegedly undermines the foundations of responsibility" (2011, p. 896). However, even here he concludes by noting that: "Solving the free will problem might have profound implications for criminal law doctrines and practices, such as blame and punishment" (2011, p. 898).

on a basic belief that we do not inhabit an incompatibilist hard determinist universe.[4] Hart, similarly, sidesteps hard determinism. In *Legal Responsibility and Excuses*,[5] he defines determinism to exclude any form of determinism that denies personal choice (1968, pp. 28–29).[6]

The nature of the universe in this respect, in particular the existence or otherwise of free will (as examined by Greene and Cohen), of capacity for rationality (as considered by Morse), of the ability to adjust behaviour (as espoused by Hart), of moral culpability (as explained by Moore) or of degrees of blameworthiness (as I advocate), cannot be determinatively resolved. As Morse wrote:

> No analysis of this problem could conceivably persuade everyone. There are no decisive, analytically incontrovertible arguments to resolve the metaphysical question of the relationship between determinism, libertarian free will and responsibility. (Morse, 2007, p. 213)

The Current Situation: Determinism v. Free Will, Responsibility, Rationality and Blame

Neuroscience may one day, as Greene and Cohen suggest, undermine societal beliefs in free will, capacity for rationality and ability to adjust behaviour. If it does, it will make our conceptions of moral culpability, responsibility and blameworthiness redundant. However, that day has not yet dawned. As Ronald Dworkin notes the case for determinism has not been proved (2011, pp. 219–252).

Even Greene and Cohen recognize this when they note that "our commitments to free will and retributivism are simply inescapable for all practical purposes" (2004, pp. 1783–1784). George Fletcher similarly takes the view that:

---

[4]For a concise statement of Morse's beliefs see his Declaration of Interest (Morse, 2006, p. 398).

[5]Originally published in Hook, S. 1958. *Determinism and Freedom* and reproduced in Hart (1968).

[6]Especially n.1.

> In order to defend the criminal law against the determinist critique, we need not introduce freighted terms like "freedom of the will." (…) The point is simply that the criminal law should express the way we live. Our culture is based on the assumption that, absent valid claims of excuse, we are accountable for what we do. If that cultural presupposition should someday prove to be empirically false, there will be far more radical changes in our way of life than those expressed in the criminal law. (2000, pp. 801–802)

Morse uses Jerry Fodor's work[7] to bolster his contention that the "folk psychological theory of personhood that the law implicitly adopts seems secure" (Morse, 2007, p. 215). This folk psychology as interpreted by Morse has at its core a belief that:

> capacity for rationality is the primary responsibility criterion and its lack is the primary excusing condition. Now, it is simply a fact about human beings that they have different capacities for rationality in general and in specific contexts. Once again, for example, young children in general have less rational capacity than adults. It is also true that rationality differences differentially affect agents' capacity to grasp and to be guided by good reason. (2007, p. 215)

Therefore, for the time being at least, the law will continue to operate based on some form of folk psychology in which terms such as responsibility, rationality, choice, moral culpability and blame are assumed to be meaningful and relevant. However, that does not mean that the cognitive sciences cannot still be relevant and that the folk psychological basis of the law cannot evolve as understandings from the cognitive sciences increasingly permeate the thinking of the population at large including judges, jurors, law makers and all those involved in the processes of the law.

---

[7] Particularly Fodor's comment "if commonsense intentional psychology were really to collapse, that would be, beyond comparison, the greatest intellectual catastrophe in the history of our species; if we're wrong about the mind, that's the wrongest we've been about anything" (Fodor, 1987, p. xii, quoted in Morse, 2007, p. 216).

Lessons from Neuroscience

Neuroscience and the cognitive sciences more generally have enriched our understanding of the way brains develop, the typical roles of different parts of the brain and given us more insight into what can go wrong with the workings of the brain. Arguably, this has often just reinforced and helped explain lessons from the behavioural sciences. This is certainly what Morse would argue (Morse, 2006, 2013). In his words "behavior is the gold standard; neurodata is simply a handmaiden" (Morse, 2013, p. 521). Later in the same article, he expresses the view that "for now, neuroscience provides little added value to legal responsibility assessments and policy beyond what behavioral science already provides" (Morse, 2013, pp. 524–525).

Morse's argument can be challenged using the case of Mohammed Sharif.[8] Sharif was charged with conspiracy to defraud. His father had made several fraudulent claims including one for compensation from the Criminal Injuries Compensation Board relating to a head injury allegedly sustained by Mohammed Sharif. It was claimed that the head injury had led to a severe deterioration in Sharif's physical and mental condition. Sharif denied any knowledge of the claim made by his father and denied being involved in the conspiracy. The question of whether Sharif was fit to plead[9] arose. Following interviews with Sharif and observation of a video alleged to be of Sharif, the experts appointed by both the prosecution and the defence concluded that Sharif was fit to stand trial. The defence expert, Dr Guly, was of the view that "it was highly improbable" that Sharif "was suffering any serious mental illness or organic brain injury".[10] Two brain scans[11] were conducted after the decision was made that Sharif was fit to stand trial, but prior to trial. Dr Forbes, a consultant neuroradiologist, found on the basis of the MRI

[8] *R v Mohammed Sharif* [2010] EWCA Crim 1709.

[9] To be fit to plead under English law a person must be able to understand the charges, decide whether to plead guilty or not guilty, exercise his right to challenge jurors, instruct his legal advisors, follow the proceedings and give evidence on his own behalf (*R v Pritchard* [1836] 7 C & P 303 and *R v M (John)* [2003] EWCA Crim 3452).

[10] *R v Mohammed Sharif* [2010] EWCA Crim 1709 [7].

[11] One Magnetic Resonance Imaging (MRI) scan and one Electroencephalogram (EEG).

"that there was enlargement of the extra cerebral spaces in the brain" indicating "mild generalized atrophy of the brain".[12] Dr Launer, on the basis of the EEG, found evidence of "conversion syndrome or longstanding functional psychosis in addition to an organic brain syndrome".[13] Nevertheless, the prosecution expert witness, Professor Deakin, maintained his view that Sharif was malingering and was fit to plead. At trial there was a further finding that Sharif was fit to plead. Sharif took no part in his trial. He was found guilty by the jury. The judge sentenced him to three years imprisonment.

Subsequently, Sharif had a further MRI scan. This showed further brain atrophy indicating a chronic degenerative disorder. Over the next few years, further experts investigated Sharif's case including a consultant physician, a professor of neurology, a professor of learning disabilities, a consultant in neuropsychiatric genetics, an expert in genetics and ophthalmology, a consultant neuroradiologist and a consultant neurologist. In total at least 12 experts had looked at Sharif's case by the time his appeal case was heard and over 11 years had elapsed since his conviction. Professor Deakin, the professor of psychiatry whose expert evidence at trial and at the earlier fitness to plead hearing had been that Sharif was malingering, gave evidence at the appeal. Professor Deakin had changed his view and was now persuaded that Sharif was suffering from a neurodegenerative condition. At trial Deakin's evidence had been influenced by Sharif's behaviour on a family video that the police had seized. Sharif's legal team had always argued that Sharif was not the person in the video, but at appeal they conceded that they could not prove that it was not Sharif. However, in the light of the second MRI scan, Professor Deakin agreed that it was likely that Sharif had been suffering from this condition at his original trial and as a result accepted that he might have been unfit to plead. In the words of Lord Justice Kay, Professor Deakin had "radically altered his opinion".[14]

This case illustrates that when, in an article criticizing extravagant claims for neuroscience, Morse states that "common sense dictates that

[12] *R v Mohammed Sharif* [2010] EWCA Crim 1709 [10].
[13] Ibid.
[14] Ibid [24].

we should believe the behavioral evidence rather than the neuroscience evidence" (2006, p. 401),[15] he is overclaiming for behavioural science. The proper way forward for the law is to make use of the best evidence available. This is often going to be based on a combination of neuroscience and the emerging cognitive sciences alongside the traditional psychiatric and behavioural sciences. Morse does recognize this. Even in his article entitled "Brain overclaim syndrome and criminal responsibility", he accepts that there could be cases where either the behavioural evidence seems clear, but the neuroscientific evidence shows that appearances are deceptive or cases where the behavioural evidence is unclear (Morse, 2006, pp. 400–401). I think his recognition of the value of neuroscientific evidence is too limited and too grudging. There is an increasing body of research into the use of neuroscientific evidence in the criminal courts[16] which indicates that neuroscience is assisting courts to achieve justice. Most of the jurisdictions studied do not have a diminished capacity defence; one exception, as noted previously, is the Netherlands. Here de Kogel and Westgeest (2015) found that of the 72 cases in which defendants used neuroscientific and/or behavioural genetic evidence to support an accountability related claim in six the court found the defendant not accountable and in 55 the court found diminished accountability. This suggests that if a similar partial defence were available in more jurisdictions neuroscientific evidence could prove valuable.

Juveniles, Responsibility and Blame

We know that children do not, in general, display the same level of maturity as adults. This is already reflected in legal systems that impose a

[15] Morse makes a similar claim in *Brain overclaim redux* when he states: "If the finding of any test or measurement of behavior is contradicted by actual behavioral evidence, then we must believe the behavioral evidence because it is more direct and probative of the law's behavioral criteria" (2013, p. 518).

[16] See for example: Alimardani (2018, 2019), Alimardani and Chin (2019), Catley and Claydon (2015), Chandler (2015), de Kogel and Westgeest (2015), Denno (2011, 2015), Farahany (2015), and Hafner (2019).

minimum age below which children cannot be held criminally responsible. However, the minimum age of criminal responsibility varies from country to country and even within countries (Child Rights International Network, 2019). There are also differences in how countries respond to criminal behaviour by juveniles with some hearing cases in special juvenile courts and some treating juveniles who commit crimes, but who are in an age range slightly above the minimum age of criminal responsibility differently with the focus instead being on education, rehabilitation and training rather than (at least in theory) punishment. However, there is no consistency—minimum ages of criminal responsibility vary from seven to 16, and some jurisdictions including some US States have no age of criminal responsibility, but instead impose a capacity test.[17] Looking worldwide there is no clear consensus as to the age that should be adopted or the criteria that should be applied. Moore considers that: "Juveniles as a class are considered incapable of committing crime because they are young and immature" (1997, p. 485). He expands on this when he states:

> the very young, lack that general capacity we call rationality. They lack the ability to form and act on valid practical syllogisms that proceed from intelligible desires and from rational beliefs and which do not self-defeatingly conflict with other desires and beliefs held by the agent. (Moore, 1997, p. 62)

Morse similarly focusses on lack of rationality (2011, pp. 936–937) as the basis on which the young are not criminally responsible. Morse argues (2006, 2013) that recent US Supreme Court decisions[18] on the punishment of those aged under 18 when they committed their offences did not focus on the evidence derived from neuroscience despite the submission of an amicus brief written on behalf of the scientific community in

[17] For information on ages of criminal responsibility and approaches to juvenile offenders see Cipriani (2009) and CRIN (2019).

[18] *Roper v Simmons*, 543 U.S. 551 (2005); *Graham v Florida* 6 130 S. Ct. 2011 [2010]; *Miller v Alabama* 567 U.S. 460 (2012).

Roper v Simmons.[19] Aliya Haider, who was involved in the team submitting the brief, takes a very different approach emphasizing the reliance on emerging neuroscientific knowledge[20] and claiming the importance of the science in the Supreme Court decision.[21]

In the UK, the issue of the age of criminal responsibility has recently been the focus of attention. In Scotland, the Scottish Parliament raised the age of criminal responsibility from eight to 12 in 2019.[22] The age of criminal responsibility for the rest of the UK is 10. In 2018, a note was produced for the UK Parliament (Parliamentary Office of Science and Technology [POST], 2018) outlining the current law, its historic development, policy considerations, arguments for reform, the state of public opinion and the UK Government's stance.

Unlike Morse, who is quite dismissive of insights from neuroscience, the Parliamentary note focusses particularly on the growing understanding of brain and behavioural development in children: "Advances in our understanding of the neurobiological processes underpinning adolescent behaviour have raised questions regarding the extent to which children should be held culpable for their actions" (POST, 2018, p. 2). The report notes not just the behavioural changes exhibited through adolescence,[23] but also gives prominence to neurodevelopmental alterations during adolescence. This approach, recognising the significance of both behavioural and neurocognitive science is, I believe, appropriate. The report notes that the prefrontal cortex, the area of the brain associated with controlling decision making, planning, social interaction and inhibiting risky behaviour, is one of the last brain areas to fully mature.

---

[19]The amicus brief in *Roper v Simmons* was submitted on behalf of the American Medical Association, the American Psychiatric Association, the American Society for Adolescent Psychiatry, the American Academy of Child and Adolescent Psychiatry, the American Academy of Psychiatry and the Law, the National Association of Social Workers, and the National Mental Health Association.

[20]"we relied on emerging scientific data for support to argue that the adolescent brain is not fully formed, and consequently, adolescent decision-making capacity and risk-taking behavior is far different than that of an adult" (Haider, 2006, p. 370).

[21]"The science brief thus played an important role in the Court's decision in *Roper*" (Haider, 2006, p. 374).

[22]The age of criminal responsibility in Scotland was raised from 8 to 12 by the Age of Criminal Responsibility (Scotland) Act 2019.

[23]Impulsive behaviour, increased risk taking and sensation seeking (POST, 2018, p. 2).

On the other hand, the amygdala and the ventral striatum (associated with risk and reward) undergo rapid development and "become hyper-responsive in adolescence" (POST, 2018, p. 2).[24] This combination of rapid development of "reward systems" alongside the slow development of the "control system" can be viewed as simply an explanation for behaviour in adolescence that had previously been observed, but I would argue that it adds to the picture and assists in understanding to what extent adolescents can control their behaviour. In examining this greater understanding that has arisen through these developments in the cognitive sciences, the report cites a number of neurodevelopmental studies and reports.[25] While not the place of a Parliamentary note to recommend law reform, it notes that there was "no obvious reason" (POST, 2018, p. 2) for the age of criminal responsibility to have been set at 10 and states the current law does not meet the UN Convention on the Rights of the Child (UNCRC) requirements that the minimum age of criminal responsibility should reflect children's intellectual, mental and emotional and immaturity and should not be set below the age of 12 (UNCRC, 2007, para 32).

Under English law, a child aged between 10 and 14 was presumed to be incapable of committing a criminal offence. This was a rebuttable presumption. To secure a conviction, the prosecution had to show that the child knew that what she was doing was seriously wrong.[26] This presumption was abolished by s.34 of the Crime and Disorder Act 1998. In *R v JTB*,[27] the House of Lords confirmed that the legislation had abolished not just the presumption, but also the associated defence stating:

---

[24] See particularly Box 2.

[25] Sources cited included: Blakemore (2018), Blakemore and Choudhury (2006), Blakemore and Mills (2014), Dillon et al. (2009), Fazel et al. (2008), Fjell et al. (2013), Gardner and Steinberg (2005), Giedd and Rapoport (2010), Lebel et al. (2012), Nickerson and Nagle (2005), Petanjek et al. (2011), Royal Society (2011), Sowell et al. (1999), Steinberg et al. (2009), Sussman et al. (2007), Tamnes et al. (2013), Van Leijenhorst et al. (2010), and Wolf et al. (2015).

[26] See for example *R v Gorrie* [1918] 83 JP 136; *JM (A Minor) v Runeckles* [1984] 79 Cr App R 255.

[27] [2009] UKHL 20.

> *doli incapax* was an anachronism. Children in the 20th Century had to go to school where they were, or were supposed to be, taught the difference between right and wrong. In the case of some offences it beggared belief to suggest that young defendants might not have appreciated that what they were doing was seriously wrong. [20] per Lord Phillips

The focus of Moore and Morse on the child's rationality as the basis for excluding infants from criminal liability, the focus of *doli incapax* on knowledge and the explanation as to why *doli incapax* was no longer needed by the highest court in the UK all ignore the insights of the cognitive sciences into neurodevelopment through adolescence. A child above the minimum age of criminal responsibility may be educated, she may know right from wrong, she may know when she has done something which is seriously wrong and she may be capable of being described as rational, but that does not mean that her judgement is not, at times, impaired. Her ability to make decisions, to plan, to resist peer pressure and to assess risk are all in a state of flux during her adolescence. To have a simple binary divide between criminally responsible and not criminally responsible does not reflect the situation. A child in England is not magically transformed on her 10th birthday or on her 12th birthday in Scotland into an adult.

A minimum age of criminal responsibility is simple, and it provides some certainty, but it does not necessarily achieve justice. As Hart commented:

> exemption by general category is a technique long known to English law; for in the case of very young children it has made no attempt to determine, as a condition of liability, the question of whether on account of their immaturity they could have understood what the law required and could have conformed to its requirements, or whether their responsibility on account of their immaturity was 'substantially impaired', but exempts them from liability for punishment if under a specified age. (1968, p. 229)

There is a need for a more nuanced approach. This is not achieved, through sentencing. A low age of criminal responsibility does not deter offending behaviour and children who perceive themselves as criminal as a result of a caution or conviction "are more likely to engage in deviant behaviours and align themselves with criminal peers" (POST, 2018, p. 4—citing Farrington & Murray, 2014). Convictions and cautions often must be disclosed when seeking employment. A juvenile who receives a custodial sentence for a category "A" offence[28] will never "clear" the conviction. In England and Wales, the PHOENIX database, which will be accessed for a "criminal record check", contains not just convictions but also reprimands, cautions, warnings, arrests, warrants, penalty notices for disorder and other information. The record will only be removed on the individual's 100th birthday. (ACRO Criminal Records Office, 2018, p. 5; Baldwin, 2019). This system needs to be reformed to recognize that it is not in the public interest to label all juvenile offenders as criminals. An alternative system is required to recognize that some young offenders are not criminally responsible, and others have diminished capacity rendering them less responsible and less blameworthy. These juveniles may be above the age of criminal responsibility but may be demonstrated to have impaired judgement and impulse control arising from age-related deficits in their cognitive development.[29]

Mental Disorder, Responsibility and Blame

Both Moore and Morse draw parallels between children and the mentally disordered emphasizing the two groups' lack of or diminished rationality (Moore, 1997, pp. 61–62, 534–535, 598; Morse 2011, pp. 936–937). Moore links this lack of rationality to moral responsibility:

> Insanity betokens a difference so fundamental that we deny moral agency to those afflicted with it. The insane, like young infants, lack one of the essential attributes of personhood namely, rationality. For this reason,

[28]There are over 1,000 Category A offences (ACRO, 2018, pp. 12–49) including consensual sexual offences, drug offences and offences arising from recklessness.

[29]Many young offenders are not just young but have additional neurocognitive issues. POST notes that "there are high rates of mental illness and substance abuse amongst children who offend. Many have learning disabilities (23-32%), communication difficulties (60-90%), and neuro-developmental disorders such as autism spectrum disorders and attention hyperactivity disorder (ADHD) (15% and 11-18% respectively)" (2018, p. 1).

> human beings who are insane are no more the proper subject of moral evaluation than are young infants, animals, or even stones. Only beings who, like most of us, are fairly good practical reasoners can be the subjects of moral norms. (1997, pp. 534–535)

However, rationality is only part of the insanity defence in English law. It is true that for the defence to apply the accused must be suffering from "a defect of reason", but that is not enough. The defect of reason must arise "from disease of the mind". Additionally and importantly, it must result in a lack of knowledge; knowledge either as to "the nature or quality of the act he was doing" or, if the accused did know the nature and quality of his act, knowledge as to whether he knew what he was doing was legally wrong.[30,31] An accused who is irrational but who, for example, knows what she is doing or knows what she is doing is legally wrong will not be viewed under English law as insane. Few defendants meet the insanity criteria and given the emphasis on legal knowledge neuroscientific evidence is largely irrelevant.[32] Annually, there are about 20–30 not guilty by reason of insanity verdicts in the Crown Courts of England and Wales (Law Commission, 2012, p. 5). To put this into context, the total number of cases annually decided in the Crown Court is over 100,000 (Sturge, 2019, p. 6). The insanity defence may be of interest to academic lawyers, but in practice, it is of little relevance. As the Anglo-American experience demonstrates: "Although it captures popular imagination, the insanity defense is raised infrequently and notoriously difficult to prove" (Farahany, 2015, p. 499).

[30]"to establish a defence on the ground of insanity it must be clearly proved that, at the time of the committing of the act the party accused was labouring under such a defect of reason, from a disease of the mind, as not to know the nature and quality of the act he was doing; or, if he did know it, that he did not know he was doing what was wrong" per Lord Tindal CJ (*M'Naghten's Case* [1843–1860] All ER Rep 229, 231).

[31]The question of whether "wrong" in the M'Naghten Rules meant morally wrong or legally wrong was decided in English law in *R v Windle* [1952] 2QB 826.

[32]Out of 204 reported cases in which neuroscientific evidence was used by those accused of criminal offences in England and Wales between 2005–2012, none were insanity pleas (Catley & Claydon, 2015).

The Need to Recognize Diminished Capacity

Only a very limited number of defendants successfully plead insanity. But this is not evidence that few defendants have cognitive or mental health problems. Official figures suggest that "90% of the prison population [of England and Wales] are mentally unwell", with at least 10% being treated for mental illness (National Audit Office (NAO), 2017, p. 7). Prisoners are much more likely to have a learning disability (7%) than the population at large (2%) (Sentencing Council, 2019, p. 2). On arrival in prison, 23% report prior contact with mental health services (NAO, 2017, p. 13). They are also (drawing on Scottish research) more likely to have been hospitalized for a head injury (24.7%) (McMillan et al., 2019).

Indeed, the Law Commission for England and Wales have recognized the problematic nature of the insanity defence (Law Commission, 2013, pp. 6–19). In doing so, they note that "While there are a great many people convicted of offences who have mental health problems and/or learning difficulties, the number who completely lack criminal responsibility as a result is small" (Law Commission, 2013, p. 19, para 1.83). Their proposed replacement of the insanity defence with a new defence of "not criminally responsible by reason of a recognized medical condition" (Law Commission, 2013) would similarly apply to very few defendants. It is therefore not the answer to the general problem, but it does provide a mechanism which I have previously argued (Catley, 2020) could be amended to produce a partial defence to cater for many cases where defendants are less blameworthy as a result of their diminished capacity. To satisfy my proposed partial defence:

> The party seeking to raise the new partial defence must adduce expert evidence that at the time of the alleged offence
>
> the defendant substantially lacked capacity:
>
> (i) rationally to form a judgment about the relevant conduct or circumstances:
> (ii) to understand the wrongfulness of what he or she is charged with having done; or

(iii) to control his or her physical acts in relation to the relevant conduct or circumstances

as a result of a qualifying recognized medical condition.

My proposal adopts the Law Commission's focus on rationality, understanding and control. It avoids the problems associated with the M'Naghten Rules' focus on knowledge and the English court's interpretation of wrongfulness as meaning legally wrong[33] instead adopting the approach proposed by the Law Commission that "the accused need only appreciate that the act was something he or she ought not to do" (2013, p. 56, para 4.22).

However, my proposal contains three alterations from the Law Commission proposal (2013, p. 193, para 10.8). Firstly, the word "partial" is inserted before "defence". Secondly, the word "substantially" replaces "wholly" in terms of the degree to which the party must lack capacity.[34] Thirdly, the Law Commission proposal would lead to a verdict of not criminally responsible.[35] Whereas for my proposed partial defence there would first be a determination as to whether the partial defence was made out; this would be by the jury in Crown Court cases. If satisfied a *guilty, but substantially lacked capacity* verdict would be returned. The judge would then pass sentence, but in doing so would have to explicitly address the capacity finding in passing sentence (Catley, 2020, pp. 200–205). If both the Law Commission's and my proposal were implemented, it would be open for example for a jury who were not satisfied that a defendant wholly lacked capacity to return a *guilty,*

---

[33] *R v Windle* and see footnote 31 above.

[34] My adoption of the word "substantially" mirrors the use of the term in the American Law Institute's Model Penal Code and the use of the term in the English partial defence of diminished responsibility (Homicide Act 1957 s2 (1) (b))—a partial defence which only applies to a murder charge. My adoption of the term stems in part from the response of expert witnesses to a presentation I gave at the Royal Society of Medicine (May 2016) on the Law Commission proposals at which the experts highlighted the difficulty of giving evidence asserting that a defendant "wholly lacked capacity" even though satisfied as an expert that the defendant's capacity was very seriously impaired.

[35] Disposal options would still be retained (Law Commission, 2013, p. 194, proposals 10 and 11).

but substantially lacked capacity verdict. Such a verdict would recognize that the defendant's capacity was impaired but not eliminated and would recognize that the defendant bore a reduced level of criminal responsibility.

Conclusion

The cognitive sciences have reinforced the message from the behavioural sciences that some people have reduced impulse control and impaired capacity for rational thought and reasoned judgement. This sometimes arises from developmental immaturity, in others it is a product of brain injury, brain abnormality or brain disease. Most of these individuals are not insane as that term is used either in legal or medical circles, but their capacity is diminished. There is a need, as I have argued throughout this chapter, for an intermediate stage between being fully criminally responsible and not being criminally responsible. As Justice Stevens noted in *Atkins v Virginia*:

> Mentally retarded persons frequently know the difference between right and wrong and are competent to stand trial. Because of their impairments, however, by definition they have diminished capacities to understand and process information, to communicate, to abstract from mistakes and learn from experience, to engage in logical reasoning, to control impulses, and to understand the reactions of others. (…) Their deficiencies do not warrant an exemption from criminal sanctions, but they do diminish their personal culpability. (…)
>
> With respect to retribution-the interest in seeing that the offender gets his "just deserts"-the severity of the appropriate punishment necessarily depends on the culpability of the offender. (2002, pp. 318–319)

Both Hart and Moore agree that such impairments should be reflected in punishment. Hart arguing that: "Justice requires that those who have special difficulties to face in keeping the law should be punished less" (1968, p. 24). Moore coming to a similar conclusion by a different route: "If one adopts the retributivist theory of punishment that I defend (…)

then the guiding purpose of criminal law is to punish those who deserve it in proportion to their desert" (1997, p. 256). Morse, I think rightly, goes a step further making a compelling case that culpability determination should be part of the "highly visible trial stage", rather than the "comparatively low visibility sentencing proceeding" (Morse, 2003, pp. 298–299). My proposal both moves culpability determination to the trial stage, but also makes its impact more visible in the sentencing stage.

In this chapter, I have made considerable use of the work of Herbert Hart, Stephen Morse and Michael Moore. I will end with three quotes—one from each author. Hart commented that:

> What is crucial is that those whom we punish should have had, when they acted, the normal capacities, physical and mental, for doing what the law requires and for abstaining from what it forbids, and a fair opportunity to exercise these capacities. Where these capacities are absent ... it is morally wrong to punish because 'he could not have helped it' or 'he could not have done otherwise' or 'he had no real choice'. (1968, p. 152)[36]

Hart is right, but his comment ignores those without normal capacities for whom it would have been very difficult to do otherwise. Morse's comment that "establishing that the defendant had a substantial mental abnormality at the time of the crime and therefore deserves mitigation is reasonably possible" (2011, p. 944) suggests that a defence of the type I advocate is feasible. Finally, Moore's comment that "Lawyers (...) cannot insulate their talk from the insights of an advancing science" (1997, p. 520) is perhaps the most important take home message.

Bibliography

ACRO Criminal Records Office. (2018, January 26). *Step-down model: Filtering of offences for certificates of convictions*, 2.1. Fareham: ACRO.

---

[36]This quotation was also referred to by Moore (1997, p. 550).

https://www.acro.police.uk/Acro/media/ACRO-Library/STEP-DOWN-MODEL-v2-1.pdf.

Alimardani, A. (2018). Neuroscience, criminal responsibility and sentencing in an Islamic country: Iran. *Journal of Law and the Biosciences, 5*(3), 724–742.

Alimardani, A. (2019). *An empirical study of the use of neuroscientific evidence in sentencing in New South Wales, Australia*. Sydney: University of New South Wales.

Alimardani, A., & Chin, J. (2019). Neurolaw in Australia: The use of neuroscience in Australian criminal proceedings. *Neuroethics, 12,* 255–270.

Allely, C. S. (2016). Prevalence and assessment of traumatic brain injury in prison inmates: A systematic PRISMA review. *Brain Injury*, 1161–1180.

Baldwin, C. W. (2019). *Protection from harm or more harm than good? A critical evaluation of the PHOENIX Police National Computer application and concurrent police compliance with applicable data protection legislation.* Sunderland: University of Sunderland.

Blakemore, S.-J. (2018). Avoiding social risk in adolescence. *Current Directions in Psychological Science, 27*(2), 116–122.

Blakemore, S.-J., & Choudhury, S. (2006). Development of the adolescent brain: Implications for executive function and social cognition. *Journal of Child Psychiatry and Psychology, 47,* 296–312.

Blakemore, S.-J., & Mills, K. L. (2014). Is adolescence a sensitive period for sociocultural processing? *Annual Review of Psychology, 65,* 187–207.

Bronson, J., & Berzofsky, M. (2017). Indicators of mental health problems reported by prison inmates, 2011–12. *Bureau of Justice Statistics.*

Catley, P. (2020). Personality change, criminal responsibility and diminished capacity. In A. Waltermann, D. Roef, J. Hage, & M. Jelicic (Eds.), *Law, science, rationality*. The Hague: Eleven International Publishing.

Catley, P., & Claydon, L. (2015). The use of neuroscientific evidence in the courtroom by those accused of criminal offenses in England and Wales. *Journal of Law and the Biosciences, 2*(3), 510–549.

Chandler, J. A. (2015). The use of neuroscientific evidence in Canadian criminal proceedings. *Journal of Law and the Biosciences, 2*(3), 550–579.

Child Rights International Network. (2019). *Minimum ages of criminal responsibility around the world*. https://archive.crin.org/en/home/ages.html.

Children's Commissioner. (2012). *Nobody made the connection: The prevalence of neurodisability in young people who offend.* https://www.childrenscommissioner.gov.uk/publication/nobody-made-the-connection/.

Cipriani, D. (2009). *Children's rights and the minimum age of criminal responsibility: A global perspective*. London: Routledge.

de Kogel, C. H., & Westgeest, E. J. M. C. (2015). Neuroscientific and behavioral genetic information in criminal cases in the Netherlands. *Journal of Law and the Biosciences, 2*(3), 580–605.

Denno, D. W. (2011). Courts' increasing consideration of behavioral genetics evidence in criminal cases: Results of longitudinal study. *Michigan State Law Review, 2011*(3), 967–1050.

Denno, D. W. (2015). The myth of the double-edged sword: An empirical study of neuroscience evidence in criminal cases. *Boston College Law Review, 56*(2), 493–552.

Dillon, D. G., Holmes, A. J., Birk, J. L., Brooks, N., Lyons-Ruth, K., & Pizzagalli, D. A. (2009). Childhood adversity is associated with left basal ganglia dysfunction during reward anticipation in adulthood. *Biological Psychiatry, 66,* 206–213.

Dworkin, R. (2011). *Justice for hedgehogs*. Cambridge, MA: Harvard University Press.

Farahany, N. A. (2015). Neuroscience and behavioral genetics in US criminal law: An empirical analysis. *Journal of Law and the Biosciences, 2*(3), 485–509.

Farrington, D. P., & Murray J. (Eds.). (2014). *Labeling theory: Empirical tests. Advances in criminological theory* (Vol. 18). London: Transaction Publishers.

Fazel, M., Langstrom, N., Grann, M., & Fazel, S. (2008). Psychopathology in adolescent and young adult criminal offenders (15–21 years) in Sweden. *Social Psychiatry and Psychiatric Epidemiology, 43,* 319. https://doi.org/10.1007/s00127-007-0295-8.

Fjell, A. M., et al. (2013). Critical ages in the life course of the adult brain: Nonlinear subcortical aging. *Neurobiology of Aging, 34,* 2239–2247.

Fletcher, G. P. (2000). *Rethinking criminal law.* Oxford: Oxford University Press.

Fodor, J. (1987). *Psychosemantics: The problem of meaning in the philosophy of the mind*. Cambridge, MA: MIT Press.

Friestad, C., & Kjelsberg, E. (2009). Drug use and mental health problems among prison inmates – Results for a nationwide-wide prison population study. *Nordic Journal of Psychiatry, 63*(3), 237–245.

Gardner, M., & Steinberg, L. (2005). Peer influence on risk taking, risk preference, and risky decision making in adolescence and adulthood: an experimental study. *Developmental Psychology, 41*(4), 625–635.

Giedd, J. N., & Rapoport, J. L. (2010). Structural MRI of pediatric brain development: What have we learned and where are we going? *Neuron, 67*(5), 728–734.

Greene, J., & Cohen, J. (2004). For the law, neuroscience changes nothing and everything. *Philosophical Transactions of the Royal Society B: Biological Sciences, 359*(1451), 1775–1785.

Hafner, M. (2019). Judging homicide defendants by their brains: An empirical study on the use of neuroscience in homicide trials in Slovenia. *Journal of Law and the Biosciences, 6*(1), 226–254.

Haider, A. (2006). Roper v. Simmons: The role of the science brief. *Ohio State Journal of Criminal Law, 3*(2), 369–378.

Hart, H. L. A. (1958). Legal responsibility and excuses. In S. Hook (Ed.), *Determinism and freedom.* New York: New York University Press.

Hart, H. L. A. (1968). *Punishment and responsibility: Essays in the philosophy of law*. Oxford: Clarendon Press.

Hook, S. (Ed.). (1958). *Determinism and freedom.* New York: New York University Press.

Institute for Government and The Chartered Institute for Public Finance & Accountancy. (2019). *Prisons Performance Tracker 2019.* https://www.instituteforgovernment.org/publication/perfromance-tracker-2019/prisons.

Law Commission for England and Wales. (2012). *Insanity and automatism; scoping paper.* London. https://s3-eu-west-2.amazonaws.com/lawcom-prod-storage-11jsxou24uy7q/uploads/2015/06/insanity_scoping.pdf.

Law Commission for England and Wales. (2013). *Insanity and automatism; scoping paper*. London. https://s3-eu-west-2.amazonaws.com/lawcom-prod-storage-11jsxou24uy7q/uploads/2015/06/insanity_discussion.pdf.

Lebel, C., Gee, M., Camicioli, R., Wieler, M., Martin, W., & Beaulieu, C. (2012). Diffusion tensor imaging of white matter tract evolution over the lifespan. *Neuroimage, 60,* 340–352.

McMillan, T. M., Graham, L., Pell, J. P., McConnachie, A., & Mackay, D. F. (2019). The lifetime prevalence of hospitalised head injury in Scottish prisons: A population study. *PLoS ONE, 14*(1), 1–10.

Meynen, G., (2016). Legal insanity and neurolaw in the Netherlands: Developments and debates. In S. Moratti & D. Patterson. *Legal Insanity and the Brain: Science, Law and European Courts*. Portland: Hart.

Moore, M. (1997). *Placing blame: A theory of criminal law.* Oxford: Clarendon Press.

Morse, S. J. (2003). Diminished rationality, diminished responsibility. *Ohio State Journal of Criminal Law, 29*(1), 289–308.

Morse, S. J. (2006). Brain overclaim syndrome and criminal responsibility: Diagnostic note. *Ohio State Journal of Criminal Law, 3*(2), 397–412.

Morse, S. J. (2007). The non-problem of free will in forensic psychiatry and psychology. *Behavioral Sciences & the Law*, 203–220.

Morse, S. J. (2011). Mental disorder and criminal law. *The Journal of Criminal Law and Criminology (1973-), 101*(3), 885–968.

Morse, S. J. (2013). Brain overclaim redux. *Law and Inequality: Journal of Theory and Practice, 31*(2), 509–534.

National Audit Office (NAO). (2017, June 29). *Mental health in prisons.* HC 42. https://www.nao.org.uk/wp-content/uploads/2017/06/Mental-health-in-prisons.pdf.

Nickerson, A. B., & Nagle, R. J. (2005). Parent and peer relationships in middle childhood and early adolescence. *Journal of Early Adolescence, 25*(2), 223–249.

Parliamentary Office of Science and Technology (POST). (2018). *Age of criminal responsibility.* POSTNote 577 https://researchbriefings.files.parliament.uk/documents/POST-PN-0577/POST-PN-0577.pdf.

Petanjek, Z., Judas, M., Simic, G., Rasin, M. R., Uylings, H. B. M., Rakic, P., & Kostovic, I. (2011, August 9). Extraordinary neoteny of synaptic spines in the human prefrontal cortex. *PNAS, 108*(32), 13281–13826.

Royal Society. (2011). *Brain waves module 4, neuroscience and the law.* https://royalsociety.org/~/media/Royal_Society_Content/policy/projects/brain-waves/Brain-Waves-4.pdf.

Schiltz, K., Witzel, J. G., Bausch-Hölterhoff, J., & Bogerts, B. (2013). High prevalence of brain pathology in violent prisoners: a qualitative CT and MRI scan study. *European Archives of Psychiatry and Clinical Neuroscience, 263,* 607–616.

Sentencing Council. (2019). *Overarching principles: Sentencing principles: Sentencing offenders with mental health conditions or disorders – consultation.* https://www.sentencingcouncil.org.uk/wp-content/uploads/Mental-Health-consultation-paper-Web.pdf.

Shiroma, E. J., Ferguson, P. L., & Pickelsimer, E. E. (2012). Prevalence of traumatic brain injury in an offender population: A meta-analysis. *Journal of Head Trauma Rehabilitation, 27*(3), E1–E10.

Shulman, E. P., et al. (2016). The dual systems model: Review, reappraisal, and reaffirmation. *Developmental Cognitive Neuroscience, 17,* 103–117.

Sowell, E. R., Thompson, P. M., Holmes, C. J., Jernigan, T. L., & Toga, A. W. (1999). In vivo evidence for post-adolescent brain maturation in frontal and striatal regions. *Nature Neuroscience, 2,* 859–861.

Steinberg, L., Graham, S., O'Brien, L., Woolard, J., Cauffman, E., & Banich, M. (2009). Age differences in future orientation and delay discounting. *Child Development, 80*(1), 28–44.

Sturge, G. (2019, December 16). *House of commons briefing paper: Court statistics for England and Wales* (Number CBP 8372). https://researchbriefings.files.parliament.uk/documents/CBP-8372/CBP-8372.pdf.

Sussman, S., Pokhrel, P., Ashmore, R. D., & Brown, B. B. (2007). Adolescent peer group identification and characteristics: A review of the literature. *Addictive Behaviors, 32*(8), 1602–1627.

Tamnes, C. K., Walhovd, K. B., Dale, A. M., Østby, Y., Grydeland, H., Richardson, G., et al. (2013). Brain development and aging: overlapping and unique patterns of change. *Neuroimage, 68*, 63–74.

Van Leijenhorst, L., Zanolie, K., Van Meel, C. S., Westenberg, P. M., Rombouts, S. A., & Crone, E. A. (2010). What motivates the adolescent? Brain regions mediating reward sensitivity across adolescence. *Cerebral Cortex, 20*(1), 61–69.

Wolf, L. K., Bazargani, N., Kilford, E. J., Dumontheil, I., & Blakemore, S.-J. (2015). The audience effect in adolescence depends on who's looking over your shoulder. *Journal of Adolescence, 43*, 5–14.

Cases

Atkins v Virginia 536 US 304 [2002].

Graham v Florida 6 130 S. Ct. 2011 [2010].

JM (A Minor) v Runeckles [1984] 79 Cr App R 255.

Miller v Alabama 567 U.S. 460 [2012].

M'Naghten's Case [1843–1860] All ER Rep 229.

R v Gorrie [1918] 83 JP 136.

R v JTB [2009] UKHL 20.

R v M (John) [2003] EWCA Crim 3452.

R v Mohammed Sharif [2010] EWCA Crim 1709.

R v Pritchard [1836] 7 C & P 303.

R v Windle [1952] 2QB 826.

Roper v Simmons 543 U.S. 551 [2005].

Legislation

Age of Criminal Responsibility (Scotland) Act 2019.
Crime and Disorder Act 1998.
Homicide Act 1957.

Treaties

United Nations Convention on the Rights of the Child, 2007.

Paul Catley is Professor of Neurolaw and until April 2021 was Head of the Open University Law School. His research focuses on the use and potential use of neuroscientific and genetic evidence in the courts and within justice systems more widely. His interests are wide ranging and include the use of neuroscientific evidence to detect memory and lies, the use of brain scanning to inform treatment and end of life decisions for patients with persistent disorders of consciousness and the appropriate approaches of the law in cases where brain impairment or brain injury may affect responsibility and/or capacity.

Coercion and Control and Excusing Murder?

Lisa Claydon

Introduction

Throughout Europe in recent years, there have been significant political pressures to recognize the experience of those whose behaviour changes as the result of being coerced and controlled.[1] In the UK, legislative initiatives such as the Modern Slavery Act, 2015 have recognized the harm caused to those who are coerced and controlled into slavery.

[1] See e.g. Anti-slavery, 'A call for European Union legislation on mandatory human rights and environmental due diligence, to prevent forced and child labour in global supply chains', May 2020.

L. Claydon (✉)
Open University Law School, The Open University, Milton Keynes, UK
e-mail: lisa.claydon@open.ac.uk

S. Ligthart et al. (eds.), *Neurolaw*, Palgrave Studies in Law, Neuroscience, and Human Behavior,
https://doi.org/10.1007/978-3-030-69277-3_4

In England[2] the Serious Crime Act 2015 s76 created a domestic abuse offence of controlling or coercive behaviour in an intimate or family relationship. This forms part of a government strategy to prevent violence against women and girls (Home Office, 2016).

The Domestic Abuse Bill that is before the UK Parliament in the 2020 session includes powers for the issuing of a Domestic Abuse Order (DAO). The court hearing the application for a DAO may impose restrictions on the perpetrator of the abuse. These restrictions include movement restrictions, electronic tagging and lie detection tests. S32(1) of the Bill provides the court may "impose any requirements that the court considers necessary" to protect the abused person from the perpetrator (UK Parliament Bills, 2020). A policy paper published at the same time makes clear the concerns that have led to the publication of the Bill,

> There are some 2.4 million victims of domestic abuse a year aged 16 to 74 (two thirds of whom are women) and more than one in ten of all offences recorded by the police are domestic abuse related. (UK Government, 2020)

One of the defining features behind the creation of the new criminal offences appears to be the acceptance that abuse by coercion and control is harmful of itself, changes the behaviour of victims and blights their lives. (Home Office, 2016)

Similarly, s1 of the Modern Slavery Act[3] criminalizes holding another person in slavery or servitude or requiring the performance of forced or compulsory labour. The Modern Slavery Act also creates a defence where a criminal act is committed by someone over the age of 18 and

---

[2]References to England or English Law include Wales.

[3]1 A person commits an offence if:

1. the person holds another person in slavery or servitude and the circumstances are such that the person knows or ought to know that the other person is held in slavery or servitude, or
2. the person requires another person to perform forced or compulsory labour and the circumstances are such that the person knows or ought to know that the other person is being required to perform forced or compulsory labour.

the person is compelled to do that act by reason of “slavery or relevant exploitation”.[4] This defence is very narrowly drafted and restricts the excuse to those who act where a “reasonable person” in the “same circumstances… having the person’s relevant characteristics would have no realistic alternative to doing that act”. Schedule 4 of the Act limits the range of criminal offences to which the statutory defence may apply. Most serious offences are excluded from the statutory defence.

The aims of the new legislation and proposed legislation are highly commendable. Such changes are noteworthy precisely because historically the criminal law has struggled to accommodate the lived experience of those who suffer violence and other forms of abuse particularly in homicide cases (Kennedy, 2018). I have previously written about the discordance between the approach of the common law to criminal defences in England and new legislative approaches to coercion and control with reference to the defence of duress. (Claydon, 2019) An interesting question is how does the law treat those who have been severely abused where that abuse results in the abused victim committing a criminal act?

Recognizing Behavioural Excuses?

The criminal law has over many years developed a distinctive approach to defining criminal liability. Responsibility for criminal acts is assessed by the courts at a particular moment, namely the precise moment the criminal act occurred. The law as interpreted and applied yields specific definitions of crimes. The personal circumstances of those who commit crime are largely ignored but may become relevant at the sentencing stage once guilt has been established (Norrie, 2005). This process effectively reduces the ability of the criminal justice system to make any substantive moral judgement as to the blameworthiness of the individual criminal conduct.

---

[4] s45.

In Anglo-American jurisprudence, there is a tendency to analyse the criminal law as being in two parts. The general part that is "general doctrines, rules and definitions" and a "special part" which defines particular "offences such as murder, rape and theft" (Duff & Green, 2005, p. 1). Duff and Green (2005, p. 1) point out that in adopting this approach the discussion of the special part tends to be removed from more general theorizing about the appropriate direction of criminal liability, even at an academic level. The fragmented nature of the criminal law has been the subject of robust critique. Particularly the law's inability to theorize responsibility outside of its normative concepts of the "Kantian individual who is autonomous, responsible for and in control of, her actions" (Norrie, 2000, p. 93). This is problematic for an accused, as an individual is "judged in isolation from the substantive moral contexts within which she acts" (Norrie, 2000, p. 94). Whereas attributions of responsibility outside the criminal justice system tend to be made in the context of the known circumstances surrounding the act for which responsibility is being attributed.

Before exploring this further I want to consider an assertion made by Greene and Cohen (2004)), *For the law, neuroscience changes nothing and everything.* To do this I want to echo a point made by Lacey in another context.

> The argument is that law, by policing its own boundaries via its substantive rules and rules of evidence, constitutes itself as self-contained, as a self-reproducing system. There is, hence, a certain "truth" to this aspect of law. But by standing back so as to cast light on the point of view from which law's truth is being constructed, we can undermine the law's claims to objectivity. (Lacey, 1998, p. 8)

Thus, law when it claims to be objective may often be critically evaluated as taking a specific viewpoint. The value of cognitive neuroscience to the law lies in its ability to add incrementally, and slowly, to our understanding of human behaviour, and illuminate the path the law has taken or, may choose to take in future. Excellent research from the cognitive brain sciences poses a challenge to more traditional legal views regarding how society ought to judge the actions of others. The explanations of human behaviour provided by cognitive science pose an interesting counterpoint to the more normative values of the criminal law. This is not to suggest that science will replace criminal law judgements. Rather to emphasize that scientific understandings already influence the development of the law by providing evidence in court. This evidence of itself provides the law with an opportunity to develop better understandings and to improve the focus of its judgement.

New understandings of human behaviour have informed the criminal justice system's approach to memory evidence (Shacter & Loftus, 2013) its understanding of bias in courtroom actors (Aono et al., 2019) and some approaches to punishment and deterrence (Greene & Cohen, 2004; Goodenough, 2004). Cognitive neuroscience is informative regarding our intuitive feeling of being less in control of our actions when coerced or controlled (Caspar et al., 2016). Perhaps the greatest value of science is in pointing out what we do not know, by emphasizing that "in order to progress we must recognize our ignorance and leave room for doubt" (Feynman, 1955, p. 4).

Thus, the claim in this chapter is not that the cognitive sciences will establish a definite line between criminal responsibility and irresponsibility. The more interesting discussion is to examine some component parts of criminal law that inform decisions of criminal responsibility and assess whether, in the light of new knowledge from the cognitive sciences about human behaviour, the normative stance the law takes is fair or justified?

Difficulties: Recognizing the Effects of Coercion and Control in Criminal Defences

A case that attracted a great deal of press and public attention in the UK was that of Sally Challen. The facts of the case are that Challen killed her husband. The circumstances surrounding the killing are complex and convoluted. It was not disputed that Challen had suffered years of abuse at the hands of her controlling husband. The issue was the relevance of this abuse to Challen's criminal responsibility for her subsequent violent behaviour in murdering her husband.

Challen's case does not stand alone. There are a number of cases where women have defended a murder charge by claiming that they were unable to retain control of their actions because of the behaviour of someone with whom they had a close relationship.[5] Where Challen's case is different, is that the appeal against conviction was based mainly on the effect of her husband's controlling and coercive behaviour on her ability to live a life without him. The argument made, on her behalf, was that it was not possible for her to walk away from the relationship. Her reaction was not one of jealousy on finding her husband was once again deceiving her regarding his sexual infidelities, or his intentions about their shared future, rather her reaction was the product of years of his coercive and controlling behaviour.

The case received wide media coverage.[6] The facts of Sally and Richard Challens' relationship are stark. Sally met Richard when she was 15. He was 22. They married when she was 25.[7] Richard's actions during the marriage caused her considerable distress. Justice for Women supported her during her appeal against conviction. The organization's website describes the relationship between Richard and Sally:

> the deceased criticising her weight, demanding that she did everything in the house (he was unwilling to make himself a cup of tea) making passive aggressive threats by withdrawing and refusing to discuss his behaviour. His behaviour involved visiting brothels, being unfaithful and doing things which were designed to humiliate her. By way of example, he had a picture taken of himself in a Ferrari surrounded by naked female models. He had the picture made into a Christmas card and sent it to mutual friends. The deceased would not allow Sally to have friends of her own or to socialise on her own. He was financially abusive spending money on himself while the money Sally earned was used to purchase necessary household items. (Justice for Women)

[5] *R v Thornton* [1992] 1 All ER 306, *R v Ahluwhalia* [1992] 4 All ER 889, *R v Hobson* [1997] Crim L R 759.

[6] After the appeal decision was released a television documentary was shown: The Case of Sally Challen, BBC2 January 3, 2020.

[7] *R v Challen (Georgina Sarah)* [2019] EWCA Crim 916 [7].

Sally left Richard and started divorce proceedings against him. However, in 2010 she decided that she was unable to live without him and the proceedings were rescinded at her request. Later, she became suspicious of his motives for agreeing to a reconciliation and started tracking his activities on Facebook and checking his mobile phone. The murder took place after she found out that Richard was meeting another woman the next day.

Sally killed Richard while he was eating. Richard was killed by "severe blows" that she inflicted using a hammer that she had taken to the house. The case report states that after the killing she covered his body with curtains and left a note which read "I love you Sally". She then went home and "spent the evening with one of her sons who did not notice anything unusual". The following morning, she drove her son to work and then drove herself to Beachy Head, where she intended to commit suicide.[8]

During negotiations to prevent her suicide, she made a number of comments about the circumstances in which the killing was committed. She told the police that Richard had told her to "treat his infidelity like a bereavement and 'get over it'".[9] She told a chaplain who approached her prior to the arrival of the police that she had killed Richard and is reported to have said—"'if I cannot have him—no one can'".[10]

The police negotiator reported that Sally said Richard would take her back if she signed a postnuptial agreement. She had agreed to this before she discovered he was seeing someone else. It was after having listened to emails and voicemail on Richard's phone—she lost control. Sally told the negotiator that she "had been treated appallingly badly" by Richard over many years. She also said "'I should be put in a padded cell somewhere, because I have gone completely off my rocker, I am just so very depressed'".[11]

At her trial for murder, the prosecution portrayed her actions as those of a jealous wife. Their forensic psychiatrist gave his opinion that Sally

---

[8] [2019] EWCA Crim 916 [11–13] – all quotations.

[9] [14].

[10] [13].

[11] [14].

was not suffering from any "mental illness or abnormality of mind at the time of the killing".[12] The defence's plea of diminished responsibility would then have required that an abnormality of mind be established.[13]

Sally, her two sons, friends of the family and a cousin gave evidence:

> David and James Challen told the jury they thought their father had behaved badly towards the appellant. They described her doing everything for him; he controlled her and decided what they would do as a couple. She had not been a happy woman for about ten years. She became particularly distressed when she discovered that the deceased had been visiting a brothel. She often referred to it and became very suspicious of the deceased and his behaviour. She frequently accused him of infidelity. The deceased refused to engage with the appellant and told her 'to get over it' and not question him about it. They knew that the appellant examined Mr Challen's text messages and emails. The deceased himself questioned whether the appellant was mentally unstable, and she began to question herself as to whether or not she was going insane.[14]

The psychiatric expert for the defence argued that Challen was suffering from a depressive disorder that amounted to an "abnormality of mind".[15] She was found guilty by the jury and sentenced to life imprisonment for murder.

[12] [20–21] quotation [21].

[13] This partial defence had been amended by the time of her appeal. This is the provision that applied at the time of the trial.

> Where a person kills or is a party to the killing of another, he shall not be convicted of murder if he was suffering from such abnormality of the mind (whether arising from a condition of arrested or retarded development of mind or any inherent causes or induced by disease or injury) as substantially impaired his mental responsibility for his acts and omissions in doing or being a party to the killing … (Homicide Act 1957 s2(1))

[14] [25].

[15] [22].

Coercion and Control as a Defence?

Challen successfully appealed against conviction for murder in 2019.[16] On appeal, her counsel distinguished the effects of coercion and control from that of "battered women's syndrome".

> Ms Wade accepted that the courts have recognised the concept of battered person syndrome, but that syndrome focuses on the psychological impact of repeated physical abuse, whereas coercive control focuses on systemic coercion, degradation and control. The lack of knowledge about the theory of coercive control at the time of the appellant's trial, meant that the partial defence of diminished responsibility was not put as fully as it could have been and the defence of provocation was not advanced at all by counsel then representing the appellant.[17, 18]

Counsel for the appellant argued, *inter alia,* that the effect of coercive control was that the person entrapped would react violently because they could perceive no other way of escaping their abuser.[19]

---

[16]The Court of Appeal quashed Challen's conviction and ordered a retrial. In their reasoning the Court of Appeal accepted that there was new evidence based on the effect of Richard's coercive and controlling behaviour on Sally. This new evidence could have strengthened a defence of diminished responsibility at her trial. The evidence needed to be tested before a jury. However, Challen pleaded guilty to manslaughter before such a trial could take place and her plea was accepted by the CPS Crown Prosecution Service. https://www.bbc.co.uk/news/uk-england-surrey-48554239.

[17][37].

[18]The Court referred to evidence from Evan Stark, the sociologist whose expertise had been recognised by the Home Office when drafting the new offence under the Serious Crime Act.

> In coercive control, abusers deploy a broad range of non-consensual, non-reciprocal tactics, over an extended period to subjugate or dominate a partner, rather than merely to hurt them physically. Compliance is achieved by making victims afraid and denying basic rights, resources and liberties without which they are not able to effectively refuse, resist or escape demands that militate against their interests [38].

[19]"In cases of coercive control the risk that one or both parties will be severely or fatally injured is a function of a victim's level of entrapment, the degree to which due to fear, violence and/or the extent of control, she has been deprived of or otherwise lacks the non-violent means effectively to resist, refuse, defend against and/or escape from demands, attacks, betrayals. In these circumstances, while the victim's vulnerability weighs the scale against her survival, the sense of having no way out can also fuel a powerful rage against the perceived source of her containment" [39].

It is difficult not to feel a great deal of sympathy for women in Sally Challen's position—it is also difficult not to be shocked at the amount of violence meted out to her victim. The pertinent question is: how is her degree of culpability to be fairly assessed?

The difficulty for the criminal law is to reflect in its normative framework on Challen's blameworthiness, and how that might reduce her liability for murder. The English criminal law has precise categories of attribution into which the action of an accused must fit to be deemed less blameworthy. The argument put on her behalf was not that she should be found not guilty of any offence. Rather it was that the blameworthiness of her conduct was reduced by the effect of Richard's coercive and controlling behaviour upon her ability to escape from the relationship.

Reform of the Partial Defences to Murder—The New Partial Defences[20]

Dissatisfaction with the way that the English criminal law was treating abused women who killed was, in part, the background to the publication of two English Law Commission reports: *Partial Defences to Murder* (Law Com No 290, 2004), and *Murder, Manslaughter and Infanticide* (Law Com No 304, 2006). The Law Commission proposed two new partial defences to murder to achieve a fairer outcome for those who killed in circumstances that made the killing less blameworthy. The old provocation defence was said to be too gendered to achieve a fair outcome.[21] The new partial defence, loss of control, aimed to provide

[20]For those unaware of the use of partial defences to murder in England and Wales these defences, if successful, operate to reduce murder to manslaughter enabling more lenient sentencing.

[21]The previous partial defence to murder as set out in Homicide Act 1957, s3.

> Where on a charge of murder there is evidence on which the jury can find that the person charged was provoked (whether by things done or by things said or by both together) to lose his self-control, the question whether the provocation was enough to make a reasonable man do as he did shall be left to be determined by the jury; and in determining that question the jury shall take into account everything both done and said according to the effect which, in their opinion, it would have on a reasonable man.

the means for women, who were acknowledged to be slower to anger and to react to abuse with a less gendered defence. The partial defence potentially excused those who had a justifiable sense of being wronged. Additionally, a new category of defendant was able to claim the defence: those who killed out of fear.[22]

The law was reformed by the Coroners and Criminal Justice Act 2009. Section 54 of that Act creates a partial defence to murder that amends s3 of the Homicide Act 1957. The defence applies where there has been an intentional unlawful killing and where the act results from a loss of self-control. The defence is limited as the loss of control must have a "qualifying trigger". The Act also requires that "a person of D's sex and age, with a normal degree of tolerance and self-restraint and in the circumstances of D, might have reacted in the same or in a similar way to D".[23] Further restrictions are placed on the circumstances that may be considered, the reference to "'the circumstances of D' is a reference to all of D's circumstances other than those whose only relevance to D's conduct is that they bear on D's general capacity for tolerance or self-restraint".[24] Killing out of revenge[25] or jealousy[26] is specifically excluded. To further complicate matters, there is a whole section of the Act that defines the qualifying trigger referred to in s54(1).[27] This section is complex, and there are some fairly insurmountable problems within

---

[22]s55(3).

[23]s54(1)c.

[24]s54(3).

[25]s54(5).

[26]s55(6)(c).

[27]s55 Meaning of "qualifying trigger"

1. This section applies for the purposes of section 54.
2. A loss of self-control had a qualifying trigger if subsection (3), (4) or (5) applies.
3. This subsection applies if D's loss of self-control was attributable to D's fear of serious violence from V against D or another identified person.
4. This subsection applies if D's loss of self-control was attributable to a thing or things done or said (or both) which—

    a. constituted circumstances of an extremely grave character, and
    b. caused D to have a justifiable sense of being seriously wronged.

5. This subsection applies if D's loss of self-control was attributable to a combination of the matters mentioned in subsections (3) and (4).
6. In determining whether a loss of self-control had a qualifying trigger—

it for a defendant such as Sally Challen. Looking at the facts of Sally Challen's case, it seems that she would have difficulty in establishing the new defence of loss of control. It would have been particularly difficult to dispel the assertions made by the prosecution that she was acting out of jealousy. In cases where ill-treatment or abuse has taken place, it will be hard to establish that emotions such as jealousy or revenge do not play a part in triggering the violent behaviour.

The policy reasons and thoughts as to who should be excluded from the use of the defence are apparent in the Law Commission report:

> We think that the objective test should apply in the case of a person responding to fear of serious violence ... Ordinarily it would not be even partially excusable for a person in fear, but not in imminent danger, to take the law into his or her own hands. We would not, for example, want a partial defence to be available to criminal gangs who choose to deal with threats of violence from rival gangs by striking first. (Law Com, 2004, para 3.112)

This paragraph makes clear the way in which the Law Commission was approaching the reform of the partial defences to murder. It seems any reform proposal had to exclude individuals who formed parts of groups that threatened law and order. This may be understandable from a political viewpoint. However, it makes it considerably more difficult to achieve justice for individual defendants who, through coercion and control, do not perceive any way of escaping from their predicament other than by using violence.

The approach adopted to the reform of the law provides support for Lacey's claim, that the adoption and analytical refinement of the norms

---

a. D's fear of serious violence is to be disregarded to the extent that it was caused by a thing which D incited to be done or said for the purpose of providing an excuse to use violence;
b. a sense of being seriously wronged by a thing done or said is not justifiable if D incited the thing to be done or said for the purpose of providing an excuse to use violence;
c. the fact that a thing done or said constituted sexual infidelity is to be disregarded.

7. In this section references to "D" and "V" are to be construed in accordance with section 54.

underlying the general principles of the criminal law provides a framework. A framework that 'keeps out of the courtroom difficult political issues' in giving examples of such issues Lacey cites the exclusion of "human motives and the substantive justification of conduct" (Lacey, 2000, p. 91).

Problems with Mitigating Responsibility for Murder

Herein lies the problem for defendants such as Sally Challen. Moreover, recognizing different levels of moral, or criminal responsibility, for killing is problematic; and complicated in English law by the inflexibility of the disposal options available to the court once a guilty verdict is reached. Murder has a fixed punishment, a whole life sentence. Thus, even where the motive for, or circumstances of, the killing might lead to a conclusion that a defendant is less blameworthy, there is no possibility of recognizing that in a reduced sentence.

The two partial defences to murder provide a means of reducing a murder verdict to manslaughter, where the killing is found by the court to be both intentional and unlawful but fits within the legal confines of the partial defences. The partial defences are very narrowly defined. Loss of control requires proof of acting out of fear, or out of a sense of being seriously wronged and is set within limiting conditions. Diminished responsibility requires proof of a relevant medical condition. Successfully pleading one of the partial defences reduces the verdict on a charge of murder to manslaughter. If the verdict is manslaughter, the sentence is discretionary.

A close reading of the two Law Commission reports suggests that the reasons for framing the partial defences as they are framed is influenced by a small number of respondents to the consultation on reform of the law. Behavioural science, though it is considered, tends to be represented by one group of experts, those who are likely to be present in the courtroom. This is natural because those who give expert evidence in the criminal courts are more likely to have an interest in responding to Law

Commission consultations concerning changes to the law. But interestingly, the views of the same experts are treated very differently in the two Law Commission Reports.

The Commission in its 2006 report justifies the need to reform the law regarding diminished responsibility, as a need to redefine the defence to enable the law to accord with psychiatric definitions "to protect the public from those mentally ill offenders who pose a continuing threat" (Law Com, 2006, para 1.49).[28] Whereas the 2004 report states that the views of the Royal College of Psychiatrists were that the creation of a new partial defence based on a distinction between the emotions of anger and fear had no basis in science, the proposed reform was said by the body representing psychiatrists to: "rest[s] upon the assumption that "anger" cannot be a justification for 'responsive violence', but 'fear' can be. However, this assumes that the two emotions of anger and fear are distinct. In medical reality they are not" (Law Com, 2004, para 3.99).[29]

---

[28] "The introduction of the partial defence of diminished responsibility in 1957 was a welcome reform. However, medical science has moved on considerably since then and the definition of diminished responsibility is now badly out of date. We are recommending an improved definition which we have drawn up with the help of the Royal College of Psychiatrists and other expert consultees. The new definition has had wide support amongst consultees. We believe that the new definition has the flexibility to accommodate future changes in diagnostic practice, whilst ensuring that the public remains well protected from those mentally disordered offenders who pose a continuing threat" (Law Com, 2006, para 1.49).

[29] "More fully [W]e would point out that the approach adopted within the document to the relationship between provocation and self-defence, with the suggestion of a new partial defence of 'excessive self-defence', is based, at least partly, upon a legal misrepresentation of psychology and physiology. Hence, one way of reading the proposal to abolish the provocation defence 'in favour' of the new partial defence of self-defence is that it rests upon the assumption that 'anger' cannot be a justification for 'responsive violence', but 'fear' can be. However, this assumes that the two emotions of anger and fear are distinct. In medical reality they are not. Physiologically anger and fear are virtually identical, whilst many mental states that accompany killing also incorporate psychologically both anger and fear. Hence, the abused woman who kills in response even to an immediate severe threat will also be driven at least partly by anger at the years of abuse meted out to her, and perhaps her children. Again, the woman who waits until the man is 'helpless' to kill him, is likely not merely to be angry but also fearful that eventually he will kill her, and/or her children, and that there is no way of preventing it other than by the death of the man (partly because her cognitions have been so distorted by the years of abuse that she does not perceive the options for escape, for example legal options, at all in the same way as an ordinary person would do). Any legal solution to the current perceived problems with partial defences to murder which rested upon the assumption that fear and anger can (even usually) be reliably distinguished must, from a medical perspective, therefore fail" (Law Com, 2004, para 3.99).

It is possible to speculate as to why the legislation embraces the recommendation of the Law Commission resting on the responses of the Royal College of Psychiatrists in respect of one partial defence, diminished responsibility but not in relation to the defence of loss of control. Why is medical science said to be determinative in framing one partial defence and ignored in the other? Possibly the difference in treatment of the expert responses really underlines the political sensitivities of law making.

Is Neuroscience Helpful in Providing an Objective Viewpoint?

Greene et al. examined the debate within modern psychology regarding traditional "developmental theories that emphasized the role of reasoning and "higher cognition" in the moral judgement of mature adults" as opposed to more recent approaches that emphasized "the role of intuitive and emotional processes in human decision making" (Greene et al., 2004, p. 389). The aetiology of human decision-making is itself a highly disputed area in the cognitive sciences. Criminal law assumes that reasoning lies behind choices that can incur criminal liability. Greene et al. argue for a more nuanced approach where some moral judgements "are driven by social-emotional responses" and others described as "'impersonal' are driven largely by 'cognitive processes'" (2004, p. 389). But here again the reasoning of the law is distinct from philosophy and science.

Anglo-American criminal law generally takes the view that responsibility rests on the choices made by individuals. It assumes a process of cognitive reasoning leads to a choice to commit the criminal act. Therefore, in cases where a plea of loss of control or, of diminished responsibility is made what the defence must establish in court is that, for some reason, the element of choice is impaired. To put it another way, there is a need to provide evidence to go before the jury that the defendant's ability to avoid committing the criminal act was substantially impaired, removed or that the action was brought about by fear or a justifiable sense of being seriously wronged, appropriately triggered.

The expert giving evidence in a case involving the new partial defences to murder has a difficult role to play. In loss of control, the jury must think it possible that the accused's fear of serious violence, or justifiable sense of being seriously wronged; affected her responsibility for the criminal act. In circumstances where "a person of D's sex and age, with a normal degree of tolerance and self-restraint and in the circumstances of D, might have reacted in the same or in a similar way to D",[30] subject to the restrictions imposed as qualifying triggers. The triggers cannot be jealousy or sexual infidelity.

In the case of diminished responsibility, the defence expert must establish the possibility that impairment of responsibility is due to an "abnormality of mental functioning". The abnormality must arise from "a recognised medical condition", "substantially impair" her ability to do one or more of the following—to understand the nature of her conduct, "form a rational judgement, or exercise self-control". The revised definition states that the "abnormality of mental function" explains the criminal conduct "where it causes, or is a significant contributory factor in causing D to carry out that conduct".[31] Thus, experts are being called to give evidence on individual behaviour. Whereas research from cognitive sciences does not normally deal with individuals but draws on datasets that are representative of communities of individuals.

Legal concepts may seem alien to those in the cognitive sciences, not only does the law seek subjective conscious choice as the foundation of criminal responsibility but it uses, with certainty, words such as "fear" as the basis for excusing conditions. There is no established agreement between cognitive neuroscientists as to what fear might be. Mobbs investigated the disagreement on how to define and investigate fear on behalf of Nature Neuroscience (Mobbs, 2019). What emerges from the report of the discussion of fear is disagreement as to the neural basis of human fear reactions. This reinforces the comments made by the Royal Society of Psychiatrists that there is insufficient medical evidence to allow a confident assertion that fear can be distinguished from anger.

---

[30] s54(1)c.

[31] s2 Homicide Act 1957 as amended.

Cognitive sciences add to our understanding in terms of what we do not know about human behaviour and that is useful. For example, as just discussed, our scientific understanding of fear provides insufficient evidential basis for a defence if proof depends on expert evidence. It is another matter if the normative values of the law are to rest on folk psychology. But the law must understand that folk psychology in this area will be based on intuition and hunch. Similarly, where psychiatrists are giving evidence in court regarding medical conditions that may found a plea of diminished responsibility, they may state that a medical disorder has a definite causative effect on behaviour at the time the crime was committed. However, most scientific discoveries are reported in the language of correlation, not that of cause and effect. What then is the value of this cognitive science for the law? It is, quite simply, that scientific conclusions and hypotheses drawn from modern cognitive sciences provide a different viewpoint to evaluate the folk psychological approaches to liability adopted by the law. As Lacey would put it, they test Law's claim to adopt an objective stance.

Problems with the Development of the Law

Lacey argues that there is a more profound issue at work in relation to the articulation of defences by the criminal law. She suggests that the problematic nature of normative values in the criminal law is exposed when looking at jurisprudential criticisms of the old partial defences to murder. The critique in the appeal courts focussed on "the characteristics to be attributed to the reasonable person, the precise nature of the requirement that a loss of self-control be 'sudden and temporary' and so on". The discussion of these concepts diverted attention from important issues such as "criminal laws proper response to situations of gross inequality of power or the justification of resort to violence in such situations" (Lacey, 2000, p. 91). The problem here is not one for science, it is a political problem, arguably created by the law's own normative structures.

Oliver Goodenough suggests a further complication. In a thoughtful essay, he ponders the discoveries of neuroscience and what flows from the discussions that seek to use neuroscientific discoveries to discredit the normative viewpoint adopted by the law. Particularly, the Anglo-American viewpoint of criminal responsibility as requiring a "psychological model of free will". He suggests that in critiquing the criminal law too much attention is paid to "free will" and the "autonomous" actor. He accepts that legal tests of criminal responsibility are based on subjective choice. But argues that the critical "psychology is that of the punisher" (Goodenough, 2004, p. 1805). Concluding that the critique by ethicists and neuroscientists who argue that acts are not freely chosen and the normative values of the criminal law in punishing those who did not really choose how to act, may be misplaced. "The law of responsibility makes much more sense if it is looked at from the strategic position of an agent deciding whether to inflict punishment on a transgressor in a context of social interaction". Goodenough argues that from the viewpoint of the person who punishes "however counterfactual the free will proposition may be in a deterministic world, it is a strategic fiction that underlies the productivity of a punishment rule, and is a fiction that may be deeply lodged in human cognition and emotional psychology." (Goodenough, 2004, p. 1805)

On that basis the ethical question becomes more one of punishment and the arguments expressed by Greene and Cohen (2004) much more about the type of punishment. Should the punishment be retribution for the criminal act or consequentialism "more interested in effective prevention than in assessing blame" (Goodenough, 2004, p. 1806)? He points out that the bias in a system based on retributivism will be towards punishing those who could not avoid acting as they did because punishment is seen to be effective in "the world of strategic interaction" where punishment is simply seen as a response to the criminal act (2004, p. 1807). Goodenough argues that in the Anglo-American justice system the perceived effectiveness of punishment may produce a bias in the form of a commitment to punish even where an act is not truly chosen.

This raises different questions: is it possible for the law to move forward and accommodate changing understandings of human behaviour? Just how interdisciplinary should law reform become—is law reform safe when left in the hands of politicians? At least, another two chapters would be required to answer these questions satisfactorily. Arguably, the normative structures of the law are open to criticism precisely because they do not recognize that some criminal behaviour results from circumstances of socio-demographic risk or adversity. Although as Tim Newburn succinctly puts it, the relationships between social disadvantage and crime are complex—"it is hard to conclude that social inequality is anything other than of central importance in understanding crime, anti-social behaviour, criminal victimization and state punishment" (Newburn, 2016, p. 338). There is no doubt that the correlation between socioeconomic status and behaviour is extremely hard to understand and to research as Farah (2017) underlines in her review of neuroscientific research into the link between behaviour and deprivation. However, new knowledge and understanding will provide a much better informed viewpoint from which to constructively critique the normative values of the law.

Robinson and Cahill argue "Most people … do not instinctively or spontaneously think that the criminal law is 'about' behavior modification; they think that it is about punishing wrongdoers" (2006, p. 16). Reinforcing this point, they write "Criminal law, in particular, plays a central role in creating and maintaining the social consensus necessary for sustaining moral norms" (2006, p. 22). Why the new legislation in England and the reduction of Sally Challen's conviction from murder to manslaughter is interesting is because of what it tells us about a new societal consensus about behaviour that results from coercion and control.

Logically, this would suggest that the views of cognitive scientists might be of some interest to lawyers when seeking to establish an understanding of criminal behaviour. The scientist can draw on a body of research evidence and on personal observation derived from years of experience. Lawyers must apply and interpret the normative values that underpin the criminal law to the circumstances of individual cases. Politicians will continue to be keen to interpret and reflect societal views of what is just in given situations. Conversations between these various actors will hopefully move the law forward.

Conclusion

Lord Neuberger when he became President of the UK Supreme Court gave a great deal of thought to the role of science in the courtroom. Anxious to avoid unnecessary and costly disputes in the courtroom he suggested that the legal and scientific communities should work more closely together. He argued that:

> The law has much to learn from science, in terms of both scientific thinking and discoveries and inventions. Scientific thinking is inevitably different from legal thinking – the idea of what constitutes proof and the role of common sense are two examples of divergence. But, given the importance of experience, logic and humanity in both spheres, legal and scientific thought have much in common as well. (Neuberger, 2016, p. 9)

He also had something to say about lawyers working with scientists suggesting that: "lawyers and scientists who learn from each other's expertise can benefit society as a whole" (Neuberger, 2016, p. 9).

Advances in understanding take time and the pressures to chase research funding in science makes collaboration with scientists difficult. But the value of understanding when it is appropriate to punish cannot be underestimated. It is important to avoid the bias to punishment, described by Goodenough. Unjust punishment impoverishes rather than enriches society and leads to a distrust for and contempt of the legal system. Those who claim that cognitive neuroscience has little value in helping us to understand more about criminal responsibility are simply wrong.

References

Books and Articles

Aono, D., Yaffe, G., & Kober, H. (2019). Neuroscientific evidence in the courtroom: A review. *Cognitive Research: Principles and Implications, 4,* 40. https://doi.org/10.1186/s41235-019-0179-y.

Caspar, E. A., Christiansen, J. F., Cleeremans, A., & Haggard, P. (2016). Coercion changes sense of agency in the human brain. *Current Biology, 26,* 585–592.

Claydon, L. (2019). Coercion changes sense of agency. In A. Walterman, D. Roef, J. Hage, & M. Jelicic (Eds.), *Law, science, and rationality* (pp. 237–261). The Hague: Eleven.

Duff, R. A., & Green, S. P. (2005). Introduction: The special part and its problems. In R. A. Duff & S. P. Green (Eds.), *Defining crimes; essays on the special part of the criminal law* (pp. 1–20). Oxford: Oxford University Press.

Farah, M. J. (2017). The neuroscience of socioeconomic status: Correlates. *Cause and Consequences, Neuron, 99,* 56–71.

Feynman, R. (1955). *The Value of Science* - Address given to the 1955 meeting of the National Academy of Sciences. http://www.faculty.umassd.edu/j.wang/feynman.pdf. Accessed May 4, 2020.

Goodenough, O. R. (2004). Responsibility and punishment: Whose mind? A response. In O. R. Goodenough & S. Zeki (Eds.), *Law and the brain.* Philosophical Transactions of the Royal Society (pp. 1805–1808) 359 (1451).

Greene, J., & Cohen, J. (2004). For the law, neuroscience changes nothing and everything. In O. R. Goodenough & S. Zeki (Eds.), *Law and the brain.* Philosophical Transactions of the Royal Society (pp. 1775–1185) 359 (1451).

Greene, J. D., Nystrom, L. E., Engell, A. D., Darley, J. M., & Cohen, J. D. (2004). The neural bases of cognitive conflict and control in moral judgment. *Neuron, 44,* 389–400.

Kennedy, H. (2018). *Eve was shamed.* London: Chatto and Windus.

Lacey, N. (1998). *Unspeakable subjects.* Oxford: Hart.

Lacey, N. (2000). General principles of the criminal law? A feminist view. In D. Nicolson & L. Bibbings (Eds.), *Feminist perspectives on criminal law* (pp. 87–100). London: Cavendish.

Mobbs, D. (2019). Viewpoints: Approaches to defining and investigating fear. *Nature Neuroscience, 22,* 1205–1216.
Newburn, T. (2016). Social disadvantage crime and punishment. In Hartley Dean & Lucinda Platt (Eds.), *Social advantage and disadvantage* (pp. 322–340). Oxford: Oxford University Press.
Norrie, A. (2000). *Punishment, responsibility and justice.* Oxford: Oxford University Press.
Norrie, A. (2005). *Law and the beautiful soul.* London: Glasshouse.
Neuberger, D. (2016). Stop needless dispute of science in the courts. *Nature, 531,* 9.
Robinson, P. H., & Cahill, M. T. (2006). *Law without justice: Why criminal law doesn't give people what they deserve.* New York: Oxford University Press.
Shacter, D. L., & Loftus, F. L. M. (2013). Memory and the law: What can cognitive neuroscience contribute? *Nature Neuroscience* (16), 119–123.

Reports

Law Commission. (2004). *Partial Defences to Murder.* Law Com No 290.
Law Commission. (2006). *Murder, Manslaughter and Infanticide.* Law Com No 304.

Government Publications

Home Office. (2016). *Ending violence against women and girls: Strategy 2016–2020.* UK Government.

Websites

BBC News. https://www.bbc.co.uk/news/uk-england-surrey-48554239. Accessed February 19, 2020.
Domestic Abuse Bill, UK Parliament, 2020. https://publications.parliament.uk/pa/bills/cbill/58-01/0096/20096.pdf.
Justice for Women. https://www.justiceforwomen.org.uk/sally-challen/. Accessed February 22, 2020.

UK Government, publications fact sheet domestic abuse bill. https://www.gov.uk/government/publications/domestic-abuse-bill-2020-factsheets/domestic-abuse-bill-2020-overarching-factsheet.

Statutes

Homicide Act 1957.
Modern Slavery Act 2015.
Serious Crime Act 2015.

Cases

R v Ahluwhalia [1992] 4 All ER 889.
R v Challen (Georgina Sarah) [2019] EWCA Crim 916.
R v Hobson [1997] Crim L R 759 (CA).
R v Thornton [1992] 1 All ER 306.

Lisa Claydon examines criminal law, with a particular interest in mental condition and other defences that are based on excusing conditions. She is actively researching the intersection between cognitive neuroscience and the criminal law. She was co-investigator on an AHRC-funded project entitled *A Sense of Agency*. This project examined neurocognitive and legal approaches to a personal sense of agency. Currently, she is researching what neuroscience may tell us about memory in the courtroom and looking at the effect of alcohol and drugs on criminal responsibility.

Reading the Sleeping Mind: Empirical and Legal Considerations

Ewout Meijer and Dave van Toor

Introduction

Technology such as functional magnetic resonance imaging (fMRI) and electroencephalography (EEG) allows for observation of the brain in action. Following the developments in these techniques, knowledge of brain function has increased tremendously over the last decades. This led to the relatively young interdisciplinary field of neurolaw that discusses the influence of new neuroscientific knowledge on criminal law (Catley & Claydon, 2015; Chandler, 2015; De Kogel & Westgeest, 2015; Alimardani & Chin, 2019; Hafner, 2019; Ligthart, 2019).

E. Meijer (✉)
Maastricht University, Maastricht, The Netherlands
e-mail: eh.meijer@maastrichtuniversity.nl

D. van Toor
Utrecht University, Utrecht, The Netherlands
e-mail: d.a.g.vantoor@uu.nl

S. Ligthart et al. (eds.), *Neurolaw*, Palgrave Studies in Law, Neuroscience, and Human Behavior,
https://doi.org/10.1007/978-3-030-69277-3_5

Within neurolaw, the use of brain imaging techniques for the detection of deception has attracted considerable attention. This chapter is about one of such tests, namely the EEG-based concealed information test (also known as memory detection). Memory detection relies on the premise that the perpetrator of a crime has memories about details of the crime stored in their brain, and that the presence of those memories (the 'guilty knowledge') can be established using some kind of experimental procedure. Legal evaluations of such procedures typically consider the voluntary nature of the test (e.g. Ligthart, 2019). In this chapter, we will discuss to what extent the legal arguments change if a memory detection test could be administered in other states of consciousness, specifically in sleeping participants.

We chose to write this chapter as a diptych, with each of the two sections written by the authors separately, based on their expertise. In the first section, Meijer describes if and how guilty knowledge can be detected from the sleeping mind. In the second section, Van Toor discusses how memory detection during sleep can be evaluated from a human rights perspective, focusing on the right to remain silent and the privilege against self-incrimination as recognized under Article 6 of the European Convention on Human Rights.

Reading the Sleeping Mind: Empirical Considerations

The Polygraph

> The innocent man, especially the nervous man, may grow as much excited on the witness stand as the criminal … his fear that he may be condemned unjustly may influence his muscles, glands and blood vessels as strongly as if he were guilty. Experimental psychology cannot wish to imitate with its subtle methods the injustice of barbarous police methods.

Hugo Munsterberg wrote these words in his seminal book 'on the witness stand' in 1908. Over 100 years later his criticism remains at the core of many debates about deception detection tools, most notable the debate surrounding the polygraph. This machine registers physiological signals associated with stress and emotion such as skin conductance, respiration, and heart rate. In the polygraph test as typically administered—the Control Question Test—deception or truth telling is inferred from differential responding to relevant questions (e.g. 'Did you murder x') and control question (e.g. 'did you ever do anything illegal'). More specifically, if the suspect shows increased physiological responding to the relevant question, the conclusion is that the suspect is deceptive, whereas if the suspect shows stronger reactions to the control questions, the conclusion is that the suspect's answers are truthful. Despite its widespread use, control question polygraph tests have been heavily criticized (e.g. National Research Council, 2003; Iacono & Ben-Shakhar, 2019), including for the flawed logic of the inference as so elegantly pointed out by Munsterberg in 1908: the polygraph cannot distinguish between the fear of a guilty suspect of getting caught and the fear of an innocent suspect of not being believed.

Memory Detection

> The real use of the experimental emotion-method is therefore so far probably confined to those cases in which it is to be found out whether a suspected person knows anything about a certain place or man or thing.

Munsterberg (1908) was—to our knowledge—also the first to describe an alternative to emotion-based tests such as the typical polygraph test. This idea was further developed by David Lykken in the late fifties and dubbed the guilty knowledge test (Lykken, 1959, 1960) and later the Concealed Information Test (CIT) or memory detection (Verschuere et al., 2011). Instead of testing for emotions associated with lying, the CIT probes the presence—or absence—of crime-relevant details that can only be known to the perpetrator. It has a multiple-choice format with one correct, and multiple incorrect but plausible options (e.g. What did the perpetrator take from the safe? [a] cash, [b] credit card [c] watch [d] ring [e] laptop). Options are selected such that they are equally plausible and indistinguishable for someone without intimate knowledge of the crime. Theory behind the CIT holds that the correct alternative will elicit an enhanced physiological response only in guilty suspects (Lykken, 1959).

Initial CIT research employed recordings of the autonomic nervous system as indicators of concealed information. A 2014 meta-analysis showed that skin conductance was both the most used and most accurate measure, followed by measures of respiration and heart rate (Meijer et al., 2014). In the late eighties/early nineties, researchers introduced a variant of the CIT that traded the skin conductance and respiration measures for measures of brain activity (Farwell & Donchin, 1986, 1991; Rosenfeld et al., 1988). Specifically, this variant used the P300 component of the electro-encephalogram.

The P300 is elicited by, among others, stimuli that are rare and meaningful, and occurs within 300–800 ms after stimulus presentation (Luck, 2005). The P300 is typically elicited using the oddball paradigm (Donchin, 1981). In this paradigm, a series of stimuli are delivered with each stimulus belonging to one of two categories. The first category contains rare and/or task-relevant stimuli, the second category common/frequent stimuli. The widely used auditory oddball variant, for example, presents the participant with a series of tones. The majority of these tones are low pitched, while occasionally a high pitch tone is presented. These high tones will elicit a more pronounced P300 because they deviate from the frequently presented high tones. If one is instructed to pay attention to the rare high tones, for example by instructing the participant to count them, P300 amplitude further increases (Polich, 1987). The latter is relevant for the CIT, as it shows that besides rarity, significance—in this case induced through task instructions—also influences P300 magnitude. In the P300-based CIT, the crime-relevant options form a rare category only for guilty participants. Hence, a pronounced P300 elicited by the correct alternatives indicates knowledge.

One of the problems with the physiological measures in a CIT is that they are sensitive to countermeasures. A skin conductance response can, for example, be elicited voluntarily by thinking about an emotional event (e.g. Haney & Euse, 1976). As a result, the outcome of a CIT can be manipulated (Ben Shakhar, 2011). Although it was initially suggested that the P300-based CIT would be immune to countermeasures as the P300 occurred too fast to be under deliberate control (Lykken, 1998), more recent research has shown that it is relatively easy to influence the magnitude of the P300 in a CIT. Rosenfeld et al. (2004) showed that any of the incorrect alternatives can be made significant by simple instructions such as 'when this stimulus is presented, imagine the experimenter slapping you in the face'. As a consequence, these now significant alternatives elicit a more pronounced P300, reducing accuracy of the test (Rosenfeld et al., 2004).

In sum, the premise of the CIT is supported by adequate empirical evidence, and laboratory studies indicate robust validity. Unsurprisingly, the CIT is generally positively evaluated by psychologists (Iacono, 2008; Ben Shakhar, 2012). It does, however, rely to a large extent on the voluntary, active participation of the suspect, and one can wonder to what extent this threatens applicability in real-life cases. One way to circumvent this problem is to present stimuli outside the awareness of the participants (see, e.g., Bowman et al., 2014; Maoz et al., 2012), or use Peter Rosenfeld's adapted complex trial protocol (Rosenfeld et al., 2008). But even these paradigms require the participants to press buttons upon stimulus presentation, meaning they still rely on the participant's active cooperation. Below, we suggest a more extreme possibility, namely applying the P300-based CIT in people who are asleep.

Memory Detection During Sleep

Even though sensory awareness is decreased during sleep, information processing still occurs to some extent. This is evidenced by a number of studies showing that auditory oddball tasks administered during sleep still elicit a P300. First, there is a line of research looking at the effects of acoustic properties of the stimuli. Cote and Campbell (1999a), for example, presented eight participants with auditory tones ranging in intensity from 0 to 100 dB. The loudest tones indeed elicited a P300 waveform during certain sleep stages. Again using eight sleepers and tones with varying loudness, the same authors showed that infrequently presented loud tones in a train of lower intensity tones elicited a P300 (Cote & Campbell, 1999b). Using a more standard oddball, Van Sweden et al. (1994) showed that infrequently presented high tones elicited a P300 in sleeping participants (see also Nielsen-Bohlman et al., 1991; Nordby et al., 1996). In sum, there is ample evidence showing that, at minimum, sensory processing takes place during sleep, and that stimuli eliciting a P300 in a waking state also do so during sleep.

But evidence of mere processing of acoustic properties of the stimulus such as pitch or loudness during sleep does not provide sufficient conditions for successful memory detection. The significance of the relevant alternatives in the CIT relies on meaning, which requires semantic processing. Kouider et al. (2014) provided compelling evidence for such processing during sleep. These authors first asked awake participants to classify words according to their meaning. Specifically, participants were instructed to classify words as either animals (e.g. 'dog') or objects (e.g. 'stamp') by pressing a button with their right or left hand. Interestingly, when different words belonging to the same categories (e.g. 'horse' and 'book') were presented after the participant fell asleep, brain activity still revealed motor preparation according to the previously learned association. These findings demonstrate that semantic processing takes place during sleep, and sleepers can still extract task-relevant information from external stimuli (see also Bastuji et al., 2002).

Finally, the most direct evidence for the potential of applying the CIT during sleep comes from a study by Perrin et al. (1999; see also Feld et al., 2010). These authors presented ten participants with audio recordings of their own name, alternated with recordings of other names during both wakefulness and during sleep. Names are known to be highly significant, have been shown to elicit large P300s (e.g. Berlad & Pratt, 1995), and are often used as stimuli in CIT studies (Meijer et al., 2014). Specifically, the participant's own name was presented against seven other names, resulting in a relative rare frequency of 12.5%. Unsurprisingly, the participants own name elicited a pronounced P300 during wakefulness. But even during sleep the participants' own name elicited a P300 that—according to the authors—was visible at the individual level, showing that the brain engages in semantic evaluation of stimuli during sleep.

In sum, there is compelling evidence that semantic information can be processed during sleep and can elicit a P300. Because P300 amplitude is typically smaller during sleep (e.g. Cote & Campbell, 1999a) results from previous CIT studies are unlikely to generalize, and the exact accuracy of a CIT during sleep would need to be empirically established (see, e.g., Meijer et al., 2017). Some practical obstacles still remain, and it would be interesting to see, for example, whether memory detection could be successful during pharmacologically induced sleep. But at minimum, it follows that memory detection during sleep should be theoretically possible.

Reading the Sleeping Mind: Legal Considerations

Introduction

During a forensic interview, a suspect can remain silent, and any attempt to coerce the suspect to talk, for example by using force, is considered a violation of his or her right to remain silent. When coerced to produce non-testimonial evidence to the authorities, such as a blood sample for DNA-analysis or to hand over their smartphone, a suspect can invoke the privilege against self-incrimination. Both above-mentioned rights are part of the larger right to a fair trial (Article 6 European Convention of Human Rights, hereafter: the Convention or the ECHR). According to the European Court of Human Rights (hereafter: the Court or the ECtHR) both rights lie at the heart of the notion of a fair trial, and protect the suspect against *inter alia* improper compulsion and that 'the prosecution in a criminal case seek to prove their case against the accused without resort to evidence obtained through methods of coercion or oppression in defiance of the will of the accused'.[1]

The distinction between statements on the one hand, and other evidence on the other hand, follows from a distinction made by the ECtHR in its case-law. Specifically, the Court developed the concept

[1] ECtHR, 17 December 1996, Saunders v UK, appl. no. 19187/91, para 68.

material that has an existence (in)dependent of the will of the suspect for this.[2] On the basis of this concept, national courts should assess whether the suspect can control the production of the evidence *solely* with his will. If that is the case, the material has an existence dependent of the will of the suspect and it receives a higher level of protection under the ECHR. In principle, the spoken word is protected in all but very limited circumstances,[3] whereas in most criminal justice systems, the suspect can be physically coerced to produce a cellular sample for DNA-analysis or to hand over the smartphone.

The question remains whether memory detection should be categorized the same as the spoken word, or whether the production of brain waves is analogous to, for example, the production of a blood sample. To answer this question, one needs to know whether the evidence produced during a sleeping CIT must be considered to exist dependent or independent of the will of the suspect. I will first describe, in more general terms, what types of evidence should be categorized as existing independent of the will of the suspect, and what types of evidence should be categorized as existing dependent of the will of the suspect.[4] This distinction is relevant because the categorization of sleeping CIT evidence as existing dependent or rather independent of the will of the suspect determines whether it is protected under the *right to remain silent*—when it exists dependent of the will—or by the *privilege against self-incrimination*—when it exists independent of the will. On the basis of this general description of the concept dependent or independent of the will, I will categorize brainwaves as existing independent of the will of the suspect, but memories as existing dependent of the will of the suspect.

[2]ECtHR, 17 December 1996, Saunders v UK, appl. no. 19187/91, para 69.

[3]As the Court accepted 'the limited nature of the inquiry which the police were authorised to undertake' under the British Road Traffic Act. See ECtHR 29 June 2007, O'Halloran & Francis v UK, appl. nos. 15809/02 and 25624/02.

[4]Detailed description can be found *inter alia* in Trechsel (2006, chapter 13) and Redmayne (2007).

Material that Exists (In)dependent of the Will of the Suspect

In normal situations, testimonial evidence may not be gathered with coercion because that violates the right to remain silent, while in some circumstances the suspect can be forced to produce other types of evidence. This differentiation in the level of legal protection follows from the distinction between the two types of evidence: on the one hand material that exists dependent of the will of the suspect (testimonial evidence), and, on the other hand, material that exists independent of the will of the suspect (other evidence). Whether, and, if so, under which circumstance coercion may be used to produce material that exists independent of the will of the suspect follows from some landmark cases of the Court.[5] In *Saunders*, the ECtHR stated in paragraph 69 that the protection under the right to remain silent and the privilege against self-incrimination 'does not extend to the use in criminal proceedings of material which may be obtained from the accused through the use of compulsory powers but which has an existence independent of the will of the suspect such as, inter alia, documents acquired pursuant to a warrant, breath, blood and urine samples and bodily tissue for the purpose of DNA testing'. However, the ECtHR never defined the concept 'existence independent of the will', but only gave the aforementioned (and some other) examples. However, using categorizations by the Court in other cases, some conclusions can be drawn about the distinction, and, consequently, what evidence is protected by right to remain silent and what evidence is protected by the privilege against self-incrimination.

Testimonial evidence has an existence dependent of the will of the suspect, because the material is only produced if the suspects willingly speaks or writes. It is therefore protected, with the Court only accepting limited inquiries as compelled testimony,[6] under Article 6 ECHR in the Member States of the Council of Europe,[7] (and, in the United States,

[5]ECtHR, 17 December 1996, Saunders v UK, appl. no. 19187/91, para 69; see also ECtHR (GC), 11 July 2006, Jalloh v Germany, appl. no. 54810/00.

[6]ECtHR 29 June 2007, O'Halloran & Francis v UK, appl. nos. 15809/02 and 25624/02.

[7]ECtHR, 17 December 1996, Saunders v UK, appl. no. 19187/91, para 69.

under the 5th Amendment [Mannheimer, 2011]). The production of drugs[8] and the production of documents[9] is in most cases independent of the will of the suspect. The suspect cannot control the existence of drugs nor the existence of already printed documents *solely* with his will. However, if it is unclear whether the documents exist at all or whether the documents are not readable without the suspect handing over additional information, the production of documents can be seen as dependent of the will of the suspect.[10] When the authorities do not have information about the existence of documents but try to coerce the suspect to hand over documents in a 'fishing expedition', the actual production of the documents can also be seen as a statement from the suspect acknowledging the existence of the documents and that they are in his possession. So, in European Human Rights case-law, there is some protection against the coerced production of documents but the extent of the protection is still somewhat uncertain (Lamberigts, 2016).

Whether or not material is dependent of the will can be ambiguous. For example, whereas Dutch regional courts considered that passwords to unlock smartphones exist dependent of the will,[11] Belgium's Higher Courts ruled that such passwords exist independent of the will of the suspect.[12] However, Dutch regional courts considered that unlike passwords, fingerprints to unlock smartphones exist independent of the will of suspects.[13] As a consequence, in the Netherlands a suspect can be coerced to give up his or her fingerprint to unlock their phone, but not their password, whereas in Belgium, they can also be coerced to give up their password.

---

[8]ECtHR (GC), 11 July 2006, Jalloh v Germany, appl. no. 54810/00.

[9]ECtHR, 25 February 1993, Funke v France, appl. no. 10828/84; ECtHR, 3 May 2001, J.B. v Switzerland, appl. no. 31827/96.

[10]ECtHR, 3 May 2001, J.B. v Switzerland, appl. no. 31827/96.

[11]Court of First Instance Noord-Holland 28 February 2019, ECLI:NL:RBNHO:2019:1568; Court of First Instance Den Haag 12 March 2018, ECLI:NL:RBDHA:2018:2983; Court of First Instance Rotterdam 14 December 2018, ECLI:NL:RBROT:2018:10283.

[12]Court of Cassation 4 February 2020, P.19.1086.N/1; Constitutional Court 20 February 2020, 28/2020.

[13]Court of First Instance Noord-Holland 28 February 2019, ECLI:NL:RBNHO:2019:1568.

The Categorization of the Sleeping Suspect's Guilty Knowledge

To answer the question whether sleeping CIT evidence is protected under Article 6 ECHR, it is necessary to examine whether the evidence gathered with the test exists dependent or independent of the will of the suspect. During a P300-based CIT, an EEG (or any other physical measurement) is recorded. I call this the 'biological trace'. From this biological trace, the presence or absence of guilty knowledge is inferred. I call the presence or absence of guilty knowledge the cognitive trace. Whether the biological and cognitive trace in a sleeping CIT are dependent or independent of the will of the suspect is evaluated in the next section.

The Categorization of the Biological Trace

Farrell (2010, p. 94) states that:

> asking a defendant for responses to questions while conducting a scan would clearly seem to violate this principle [the privilege against self-incrimination, DvT], the answer is less obvious in a situation where a brain scan tracks subconscious or passive perceptions to photos or statements but the defendant remains silent.

As discussed in Section "Reading the Sleeping Mind: Empirical Considerations", the CIT investigates whether a person has guilty knowledge by making a comparison between brain activity to different categories of stimuli. It therefore corresponds to the second part of Farrell's comment (*to track subconscious or passive perceptions but the defendant remains silent*). In a typical CIT, the task that the person performs involves not only listening to multiple-choice questions, but also pressing buttons to each or some of the answer options, requiring active participation (see Farwell & Donchin, 1991; Rosenfeld et al., 2008 for the procedural details). In a sleeping CIT, the person under investigation does not have to perform any physical action. The subject does not need to speak or otherwise cooperate wilfully during the investigation. S/he has to process the presented stimuli while brain activity or any other physiological response is measured. Whether the stimulus has special meaning for the suspect is reflected by his or her brain activity or physiological response. Especially during sleep, the suspect has no ability to suppress or change the response. Because the person cannot control or stop the physiological response after recognition—while *sleeping* the suspect is totally unaware that his responses are being measured—the only logical conclusion is that the biological trace collected with the sleeping CIT exists independently of the will of the suspect. In this sense, brain waves and other physiological responses can be considered 'real' physical evidence just like blood, hair, and cells.[14] Hence, the biological trace exists independently of the will of the suspect.[15]

The Categorization of the Cognitive Trace

In addition to the biological trace, the authorities also obtain a cognitive trace—whether or not guilty knowledge is present—through the administration of a sleeping CIT and the subsequent analysis of the data. Yielding this cognitive trace without consent is most problematic in the light of the right to remain silent and the privilege against self-incrimination. Both with the cognitive trace and with the spoken word, the contents of the mind are obtained as responses to the questions asked by the authorities. However, the cognitive trace is not exactly the same as the spoken word because—contrary to speaking persons—sleeping suspects are unaware about the production of information from a CIT. These similarities and differences between the spoken word and production of brain waves make it difficult to categorize the cognitive trace, *prima facie*, as being completely independent or completely dependent of the will of the suspect. Therefore, the first question to be answered is how this cognitive trace should be categorized, so that it subsequently can be assessed whether it will be protected under the right to remain silent or the privilege against self-incrimination.

[14]ECtHR, 17 December 1996, Saunders v UK, appl. no. 19187/91, para 69.

[15]The question whether it is in fact possible to conduct a coerced sleeping CIT on a not cooperating suspect exceeds the present chapter.

As starting point for the categorization of the cognitive trace, I review some positions other authors posed in the literature. Farahany (2012) compares stored memories in a mental 'safe' with prepared and stored documents in a real safe. In this sense, the information from the mental safe should be protected in the same way as documents that lie in a real safe (which could be protected under the 4th, not under the 5th, Amendment in American law, and under the privilege against self-incrimination, but not under the right to remain silent, in European Human Rights law). Easton (1998) points to the fact that the distinction between a biological trace and cognitive trace is based on a Cartesian dualism between mind and body, while that dichotomy is—according to him—pertinently incorrect. Accordingly, Easton's biocognitive trace can be considered as a 'simple' bodily response and will therefore enjoy similar legal protection under the privilege against self-incrimination compared to bodily tissue, hair, blood *et cetera*. Other authors point to the *content* of the cognitive trace, making a comparison with a statement (Farrel, 2010; Fox, 2009). Because the CIT allows for drawing the conclusion that guilty knowledge is present in the suspect's memory—a conclusion that, normally, can only be drawn if the suspect makes a statement—the analysis of the expert witness of the physiologically response should be treated as testimonial evidence (due to the content of the information).

The question remains which analogy, based on the positions in the literature discussed above, would be the most logical one: (a) memories are the same as documents, and can they be read the same as documents (cf. Farahany, 2012); (b) memories are a 'simple' product of the body, just as hair and blood (cf. Easton, 1998); or (c) memories are the same as the spoken word because the content of both pieces of evidence are the same (cf. Farrel, 2010; Fox, 2009).

The question of whether the cognitive trace exists dependent or independent of the will of the suspect is not easy to answer, because the above-mentioned three analogies can be reduced to two competing, mutually exclusive views. The cognitive trace as such is not obtained from the administration of the CIT. Only by analysing the biological trace, expert witnesses can conclude about the presence or absence of guilty knowledge. This conclusion is therefore an indirect result of the CIT, a result of analysis of a biological trace, and I concluded above (in line with the case-law on other bodily material) that this biological trace exists independently of the will of the suspect, especially during sleep. Because the cognitive trace is not obtained as such, but only through analysis of the biological trace, it can be concluded that the cognitive trace exists (at least in part) independently of the will of the suspect. The expert witnesses obtain the brain activity or physiological response independently of the will of the suspect, and the analysis of the suspect's responses is also independent of his will as well.

In a concurrent way, the acquisition of the information that is obtained can be emphasized. Guilty knowledge is by definition knowledge that only the offender possesses. That a suspect is aware of this knowledge is information that cannot be obtained in any other way than by the suspect making a statement, so the acquisition is always dependent on the will of the suspect. It is often the behaviour at the crime scene—for example which weapon the suspect used, or the location where the body was disposed of—that is prominent in CITs, and that information can normally only be gathered through a forensic interview.

If the suspect is confronted with the murder weapon during an interrogation situation, s/he can declare to be unfamiliar with the weapon or to abstain from any statement. The suspect does not have this choice when s/he undergoes a sleeping CIT. When the sleeping CIT is introduced in the criminal justice system, the right to remain silent is, at least partially, circumvented. During an interrogation, it is no longer necessary to assess whether the suspect has guilty knowledge, the authorities can obtain a CIT from the sleeping suspect and obtain the same information. By making this comparison and by emphasizing that memory detection, albeit via a detour, gains insight into memories, it can also be argued that the acquisition of the cognitive trace exists dependently on the will of the suspect.

In my view, the analysis that emphasizes the *content* of the information is most logical and therefore most persuasive. Any other categorization creates an artificial dichotomy between previously undisclosed personal information or memories made as statements and undisclosed personal information obtained through memory detection (and contrary to the will of the suspect). If memories are not also categorized as evidence that exists depending on the will of the suspect, the right to remain silent is eroded. The right to remain silent makes it possible for the suspect to remain silent during the interrogation. But if the investigating authorities want to circumvent this right by making use of memory detection, then that is possible (given the state of the art). So, if questions are asked in a CIT format, it is no longer possible for the sleeping suspect to determine that he wants to remain silent, whereas normally the contents of the mind are only revealed if the suspect decides so. Because of the possibility to circumvent the will about remaining silent or making a statement, silence is no longer an actual and effective option. By considering that every acquisition of personal information that has not previously been disclosed (e.g. in an email or diary) is the equivalent of making a statement, the value of the right to remain silent and thus also the autonomous choice in the process remains intact. I therefore argue that the cognitive trace should be labelled as being dependent on the will of the suspect, thereby obtaining the same protection as the spoken word.

The content of the information—a verbal statement made by the suspect and the interference of presence of guilty knowledge after a sleeping CIT—are the same, namely that (1) the same type of information is acquired (2) by asking the same questions albeit in a different format, (3) about information the suspect normally can determine autonomously to share that information based on his preferences. In essence, the cognitive trace is *testimonial*, because it discloses the content of memories and in a criminal procedure, the suspect can and must be able to exercise control over his thoughts as s/he is an autonomous party in the procedure and the prosecution should prove their case without evidence obtained through methods of coercion in defiance of the will of the accused. By considering the cognitive trace as being dependent of the will of the suspect, almost any form of coercion for obtaining it is unlawful, and therefore, sleeping suspects cannot be subjected to a CIT.

Conclusion

With the CIT, the memory of a suspect can be investigated indirectly to determine whether a person has guilty knowledge. Typically, application of the CIT relies on the cooperation of the suspect. There are reasons to assume that the CIT could, at least in theory, also be administered during sleep. For the criminal justice system, coerced administration can provide valuable information, especially from suspects who invoke the right to remain silent or the privilege against self-incrimination. From a human rights perspective, the coerced administration of the CIT must therefore be assessed under the right to remain silent and the privilege against self-incrimination (although an assessment of the administration under the right to respect for privacy and the right to respect for human dignity are also important).

I conclude that administration of the CIT in sleeping subjects violates the right to remain silent. Because a dichotomy between non-revealed personal information obtained by (1) a statement or in (2) another way is artificial, I believe that the cognitive trace should be considered as an equivalent of verbal statements. This means that the same rights must apply as those that apply to making a statement. Therefore, the authorities must abstain from any nature and degree of coercion that influences the decision of the suspect to make a statement. In my opinion, memory detection during sleep constitutes unlawfully circumventing the right to remain silent in order to obtain material that exists dependently of the will of the suspect.

References

Alimardani, A., & Chin, J. (2019). Neurolaw in Australia: The use of neuroscience in Australian criminal proceedings. *Neuroethics, 12,* 255–270. https://doi.org/10.1007/s12152-018-09395-z.

Bastuji, H., Perrin, F., & Garcia-Larrea, L. (2002). Semantic analysis of auditory input during sleep: Studies with event related potentials. *International Journal of Psychophysiology, 46,* 243–255.

Ben-Shakhar, G. (2011). Countermeasures. In B. Verschuere, G. Ben-Shakhar, & E. Meijer (Eds.), *Memory detection: Theory and application of the Concealed Information Test*. Cambridge, UK: Cambridge University Press.

Ben-Shakhar, G. (2012). Current research and potential application of the concealed information test: An overview. *Frontiers in Psychology, 3*(342). https://doi.org/10.3389/fpsyg.2012.00342.

Berlad, I., & Pratt, H. (1995). P300 in response to the subject's own name. *Electroencephalography and Clinical Neurophysiology/Evoked Potentials Section, 96,* 472–474.

Bowman, H., Filetti, M., Alsufyani, A., Janssen, D., & Su, L. (2014). Countering countermeasures: Detecting identity lies by detecting conscious breakthrough. *PLoS ONE, 9,* e90595. https://doi.org/10.1371/journal.pone.0090595.

Catley, P., & Claydon, L. (2015). The use of neuroscientific evidence in the courtroom by those accused of criminal offenses in England and Wales. *Journal of Law and the Biosciences, 2*(3), 510–549.

Chandler, J. A. (2015). The use of neuroscientific evidence in Canadian criminal proceedings. *Journal of Law and the Biosciences, 2*(3), 550–579.

Cote, K. A., & Campbell, K. B. (1999a). The effects of varying stimulus intensity on P300 during REM sleep. *NeuroReport, 10,* 2313–2318.

Cote, K. A., & Campbell, K. B. (1999b). P300 to high intensity stimuli during REM sleep. *Clinical Neurophysiology, 110,* 1345–1350.

de Kogel, C. H., & Westgeest, E. J. M. C. (2015). Neuroscientific and behavioral genetic information in criminal cases in the Netherlands. *Journal of Law and the Biosciences, 2*(3), 580–605.

Donchin, E. (1981). Surprise! … surprise? *Psychophysiology, 18,* 493–513. https://doi.org/10.1111/j.1469-8986.1981.tb01815.x.

Easton, S. (1998). *The case for the right to silence.* Burlington: VT: Ashgate Publishing.

Farahany, N. A. (2011). Searching secrets. *University of Pennsylvania Law Review, 160*(6), 1239–1308.

Farahany, N. A. (2012). Incriminating thoughts. *Stanford Law Review, 64*(2), 351–408.

Farrell, B. (2010). Can't get you out of my head: The human rights implications of using brain scans as criminal evidence. *Interdisciplinary Journal of Human Rights Law, 4*(1), 89–95.

Farwell, L. A., & Donchin, E. (1986). The "brain detector": P300 in the detection of deception. *Psychophysiology, 23,* 434.

Farwell, L. A., & Donchin, E. (1991). The truth will out: Interrogative polygraphy ("lie detection") with event-related potentials. *Psychophysiology, 28,* 531–547. https://doi.org/10.1111/j.1469-8986.1991.tb01990.x.

Feld, G. B., Specht, M., & Gamer, M. (2010). Differential electrodermal and phasic heart rate responses to personally relevant information: Comparing sleep and wakefulness. *Sleep and Biological Rhythms, 8*(1), 72–78.

Fox, D. (2009). The right to silence as protecting mental control: Forensic neuroscience and "the spirit and history of the fight amendment". *Akron Law Review, 42*(3), 763–801.

Hafner, M. (2019). Judging homicide defendants by their brains: An empirical study on the use of neuroscience in homicide trials in Slovenia. *Journal of Law and the Biosciences, 6*(1), 226–254.

Haney, J. N., & Euse, F. J. (1976). Skin conductance and heart rate responses to neutral, positive, and negative imagery: Implications for convert behavior therapy procedures. *Behavior Therapy, 7,* 494–503.

Iacono, W. G. (2008). The forensic application of "brain fingerprinting:" Why scientists should encourage the use of P300 memory detection methods. *American Journal of Bioethics, 8,* 30–32.

Iacono, W. G., & Ben-Shakhar, G. (2019). Current status of forensic lie detection with the comparison question technique: An update of the 2003 National Academy of Sciences report on polygraph testing. *Law and Human Behavior, 43,* 86–98.

Kouider, S., Andrillon, T., Barbosa, L. S., Goupil, L., & Bekinschtein, T. A. (2014). Inducing task-relevant responses to speech in the sleeping brain. *Current Biology, 24*(18), 2208–2214.

Lamberigts, S. (2016). The privilege against self-incrimination - A chameleon of criminal procedure. *New Journal of European Criminal Law, 7*(4), 418–438.

Ligthart, S. (2019). Coercive neuroimaging, criminal law, and privacy: A European perspective. *Journal of Law and the Biosciences, 6*(1), 289–309.

Luck, S. J. (2005). *An introduction to event-related potentials and their neural origins*. Cambridge, MA: MIT Press.

Lykken, D. T. (1959). The GSR in the detection of guilt. *Journal of Applied Psychology, 43,* 385–388. https://doi.org/10.1037/h0046060.

Lykken, D. T. (1960). The validity of the guilty knowledge technique: The effects of faking. *Journal of Applied Psychology, 44,* 258–262. https://doi.org/10.1037/h0044413.

Lykken, D. T. (1998). *A tremor in the blood: Uses and abuses of the lie detector*. Berlin, Germany: Plenum Press.

Mannheimer, M. (2011). Toward a unified theory of testimonial evidence under the fifth and sixth amendments. *Temple Law Review, 80,* 1135–1202.

Maoz, K., Breska, A., & Ben-Shakhar, G. (2012). Orienting response elicitation by personally significant information under subliminal stimulus presentation: A demonstration using the Concealed Information Test. *Psychophysiology, 49,* 1610–1617. https://doi.org/10.1111/j.1469-8986.2012.01470.x.

Meijer, E. H., klein Selle, N., Elber, L., & Ben-Shakhar, G. (2014). Memory detection with the Concealed Information Test: A meta analysis of skin conductance, respiration, heart rate, and P300 data. *Psychophysiology, 51,* 879–904. https://doi.org/10.1111/psyp.12239.

Meijer, E. H., Koch, M., & Held, K. (2017). Detecting Concealed Information during sleep. *Psychophysiology, 54,* S13.

Munsterberg, H. (1908). *On the witness stand.* New York, NY: The McClure Company.

National Research Council. (2003). *The polygraph and lie detection.* Committee to Review the Scientific Evidence on the Polygraph. Division of Behavioral and Social Sciences and Education. Washington, DC: The National Academies Press.

Nielsen-Bohlman, L., Knight, R. T., Woods, D. L., & Woodward, K. (1991). Differential auditory processing continues during sleep. *Electroencephalography and Clinical Neurophysiology, 79,* 281–290.

Nordby, H., Hugdahl, K., Stickgold, R., Bronnick, K. S., & Hobson, J. A. (1996). Event-related potentials (ERPs) to deviant auditory stimuli during sleep and waking. *Neuroreport: An International Journal for the Rapid Communication of Research in Neuroscience, 7,* 1082–1086.

Perrin, F., García-Larrea, L., Mauguière, F., & Bastuji, H. (1999). A differential brain response to the subject's own name persists during sleep. *Clinical neurophysiology, 110,* 2153–2164.

Polich, J. (1987). Response mode and P300 from auditory stimuli. *Biological Psychology, 25*(1), 61–71.

Redmayne, M. (2007). Rethinking the privilege against self-incrimination. *Oxford Journal of Legal Studies, 27*(2), 209–232.

Rosenfeld, J. P., Cantwell, B., Nasman, V. T., Wojdac, V., Ivanov, S., & Mazzeri, L. (1988). A modified, event-related potential-based guilty knowledge test. *International Journal of Neuroscience, 42,* 157–161.

Rosenfeld, J. P., Labkovsky, E., Winograd, M., Lui, M. A., Vandenboom, C., & Chedid, E. (2008). The Complex Trial Protocol (CTP): A new countermeasure-resistant, accurate, P300-based method for detection of concealed information. *Psychophysiology, 45,* 906–919. https://doi.org/10.1111/j.1469-8986.2008.00708.x.

Rosenfeld, J. P., Soskins, M., Bosh, G., & Rayan, A. (2004). Simple, effective countermeasures to P300-based tests of detection of concealed information. *Psychophysiology, 41,* 205–219. https://doi.org/10.1111/j.14698986.2004.00158.x.

Trechsel, S. (2006). *Human rights in criminal proceedings*. Oxford: Oxford University Press. https://doi.org/10.1093/acprof:oso/9780199271207.001.0001.

Van Sweden, B., Van Dijk, J. G., & Caekebeke, J. F. V. (1994). Auditory information processing in sleep: late cortical potentials in an oddball paradigm. *Neuropsychobiology, 29,* 152–156.

Verschuere, B., Ben-Shakhar, G., & Meijer, E. (Eds.). (2011). *Memory detection: Theory and application of the Concealed Information Test*. Cambridge, UK: Cambridge University Press.

Ewout Meijer is an Assistant Professor at the Faculty of Psychology and Neuroscience. He obtained his PhD in 2008 with a dissertation on the use of psychophysiological measures in lie and memory detection. He has published about a variety of topics, including deception detection, investigative interviewing, and cheating behaviour. He served as a Research Fellow at the Hebrew University of Jerusalem in 2011–2012, and as a fellow of the Israel Institute for Advanced Studies in 2020–2021.

Dave van Toor joined the Department of Criminal of the Radboud University Nijmegen in 2008. He started at the Universität Bielefeld in September 2014, first as a Research Assistant, later as Research Associate at the Department of Criminal (Procedural) Law & Criminology. In the meantime, he completed his doctoral dissertation on the legal implications of using coerced neuroscientific memory detection in criminal cases. Currently, he is Assistant Professor in Criminal (Procedural) Law at Utrecht University. His main research interest lies in the legitimacy of police investigations in view of human rights.

'Brain-Reading' in Criminal Justice and Forensic Psychiatry: Towards an Integrative Legal-Ethical Approach

Sjors Ligthart, Tijs Kooijmans, and Gerben Meynen

Introduction

One of the central issues in the present debate on neuroscience, justice and security focusses on the possibilities that 'brain-reading' offers for criminal law (e.g. Meynen, 2017; Ligthart, 2019; Pardo & Patterson, 2015; Morse & Roskies, 2013; Simpson, 2012). Scientific developments in neuroimaging, such as functional magnetic resonance

S. Ligthart (✉) · T. Kooijmans
Tilburg University, Tilburg, The Netherlands
e-mail: S.L.T.J.Ligthart@tilburguniversity.edu

T. Kooijmans
e-mail: T.Kooijmans@tilburguniversity.edu

G. Meynen
Utrecht University, Utrecht, The Netherlands

Ethics and psychiatry, VU Amsterdam, Amsterdam, The Netherlands
e-mail: g.meynen@uu.nl

S. Ligthart et al. (eds.), *Neurolaw*, Palgrave Studies in Law, Neuroscience, and Human Behavior,
https://doi.org/10.1007/978-3-030-69277-3_6

imaging (fMRI), electroencephalography (EEG) and computed tomography (CT), enable researchers to detect specific brain functions or structures which, to some extent, allow for drawing inferences about certain mental states, neural anomalies or brain features. In the light of these developments, it has been argued that brain-reading technologies can potentially contribute to the field of forensic psychiatry, e.g. to assessments of legal insanity, or predicting future dangerousness (Meynen, 2017, 2018a, 2019; Simpson, 2012). In fact, some brain-reading applications have already been deployed in criminal justice systems, most prominently to assess brain anomalies in the context of the insanity defence (Meynen, 2020; Hafner, 2019; Catley & Claydon, 2015; de Kogel & Westgeest, 2015; Farahany, 2015; Alimardani & Chin, 2019).

Whereas brain-reading technologies could, in principle, strengthen forensic psychiatric evaluations, deploying brain-reading in this context also raises fundamental, interwoven ethical and legal questions (Meynen, 2017, 2019; Richmond et al., 2012; Simpson, 2012). Although both in (medical) ethics and the law similar questions arise in this respect, the legal and ethical debates are as yet somewhat separated from each other. In line with a recent call for more collaboration between ethicists and lawyers in the neuro-legal-ethical debate (Ligthart et al., 2019), this chapter aims to provide some further direction on how ethics and the law, focussing on European human rights law, could learn from each other in the discussion on forensic brain-reading.

Such proactive collaboration between ethics and the law is especially relevant at the level of human rights, which are arguably closely related to moral rights.[1] For example, as Cruft, Liao and Renzo note regarding the philosophical foundations of human rights, one view is that human rights are moral rights that all humans possess at all times and in all places, simply in virtue of being a human (Cruft et al., 2015, p. 4). As to the European Charter of Fundamental Rights, the European Union seems to follow a similar approach, stating that the Charter 'establishes ethical principles and rights for EU citizens and residents that relate to dignity, liberty, equality, solidarity, citizenship and justice. In addition to

[1]Within the context of this chapter, we will not address the relationship between ethics and law more generally; we focus on human rights, where a close connection exists between legal and ethical principles and values, as well as scholarly discussions about them.

protecting civil and political rights, it covers workers' social rights, data protection, bioethics and the right to good administration'.[2]

The outline of this chapter is as follows. First, we briefly discuss some of the possibilities that brain-reading technologies offer in the context of forensic psychiatry (section "The Relevance of Brain-Reading for Forensic Psychiatry and Criminal Proceedings"). Next, we discuss three ethical and legal issues that arise regarding forensic brain-reading and explore how an integrative legal-ethical approach could contribute to advancing the debate on neuroscience, justice and security (section "Central Issues in the Ethical and Legal Debate: Learning from Each Other"). Finally, we make some concluding remarks (section "Conclusion").

The Relevance of Brain-Reading for Forensic Psychiatry and Criminal Proceedings

Brain-reading is defined broadly in this paper as referring to 'mind-reading procedures that rely on brain-derived data'.[3] This means that all types of neuroimaging that make it possible to say something about a person's mind can be considered 'brain-reading'. For instance, if a brain scan shows a significantly diminished volume of the hippocampus, this may tell us something about the person's memory in the sense of the capacity to retain and recall information. We may thus obtain some important information about a person's mind. This chapter is not only about technologies that try to identify pathological changes, however, but also about techniques that aim at detecting subjective states or specific knowledge, such as fMRI lie-detection and the electrophysiological detection of recognition (see below).

---

[2]See https://eur-lex.europa.eu/summary/glossary/charter_fundamental_rights.html.

[3]See Meynen (2018b). In that paper, the term brain-based mind reading is used, whereas we use 'brain-reading'. See this paper for more theoretical aspects of brain-reading. See Meynen (2017 and 2019) for some of the topics of this chapter.

Such brain-reading technologies can be of interest in the field of forensic psychiatry and criminal justice. After all, in both forensic psychiatry[4] and criminal justice, *subjective* states of the defendant or convicted offender are often essential in answering central questions of criminal law. For example, does the defendant *know* something about the crime? Does he *lie* about his alibi? Did he indeed *hear* commanding voices during the time of the offence? And how does he *feel* in particular stressful situations? Hence, forensic psychiatry and criminal justice are two areas in which subjective mental states play a central role. Importantly, in these areas, defendants and convicted offenders may well be reluctant to disclose information about such mental states—a reluctance that may manifest in lying, faking or invoking the right to remain silent. As a consequence, measures that enable the more objective assessment of the person's subjective mental states could, in principle, provide a very helpful tool to forensic psychiatrists, judges, public prosecutors and lawyers. Brain-reading could potentially become such a helpful tool in this context (Meynen, 2017; Ligthart, 2019).

Brain-reading techniques can be categorized in different ways, for instance by the technology (e.g. MRI, fMRI and EEG) or by the type of information they may yield (e.g. whether they detect particular memories, thoughts, emotions, intentions, *et cetera*). For the purpose of both an ethical and legal analysis, we distinguish between two basic characteristics: first, whether the subject has to cooperate in order to obtain the relevant information and second, whether the subject is aware of the presence and nature of the information that the brain-reading test aims to obtain. We use this categorization in part because brain-reading is as yet largely 'neuroscience fiction'. This means that the technologies—if any—that may eventually be ready for courtroom use, can be very different from those techniques used today. In order for our analysis to be relevant also for possible future technologies, we do not distinguish between different types of brain-reading based on current technological features, but rather on more general characteristics concerning the

[4]We focus in this chapter on the assessment of defendants, but another important task of forensic psychiatry concerns treatment of forensic psychiatric patients, see section "Trust".

Table 1 Two salient distinctions: knowledge/ignorance and cooperation/non-cooperation

	Cooperation required	No cooperation required
The subject knows/may know the information	– fMRI lie-detection (Farah et al., 2014) – P300 recognition (Meijer et al., 2016)	– real-time mind reading (futuristic)
The subject is ignorant about the information (need not know it for the technology to work)	– fMRI neuroprediction of rearrest (Aharoni et al., 2013) – SPECT neuroprediciton of recidivism (Delfin et al., 2019)	– MRI and CT to detect brain anomalies (Simpson, 2012)

subject of the test and the information to be obtained. The conceptual framework that is used is presented in Table 1.[5]

Let us first look at the relevance of the distinction 'knows versus ignorant'. For instance, fMRI lie-detection aims at obtaining information that the subject actually knows about: a person *knows* whether he or she is lying. Lying is something that one, by definition, has to be aware of in order for it to be a lie: one *deliberately* tells an untruth or falsehood (Pardo & Patterson, 2015, 106). The same is true for P300 measurements that try to determine whether a person recognizes an object: in general, a person *knows* when he recognizes an object or, e.g. one's own dog. This is, of course, all assuming that these technologies—fMRI lie-detection and P300-recognition tests—work well. Because the subject knows what the examiner wants to know through brain-reading technology, in principle, the subject could also have *told* the examiner what the examiner wants to know. But, in the standard case of forensic use of such technologies, the point will be that either the subject (for instance, the defendant) does not want to share the information or is considered not trustworthy. Now suppose that a forensic psychiatric patient answers all the psychiatrist's questions and then the examiner asks: do you agree with brain-reading for lie-detection? Then, the subject's response could

[5]See Meynen (2017) for roughly similar—although somewhat different—distinctions and categorizations, based on the same rationale. See also Meynen (2019).

well be: 'Why? I just gave you all the answers, don't you trust me?' Although some distrust might objectively be justified in this context, it may be increased by using such a technique and negatively affect the psychiatrist–patient relationship.

This is different if the subject does not know the information. For instance, Aharoni et al. (2013) and Delfin et al. (2019) tried to predict the risk of, respectively, rearrest and recidivism with the use of brain-reading. An offender can be considered, at least in principle, ignorant about his chances of rearrest (Meynen, 2017). One offender may solemnly believe that he will never reoffend, but his risk may be very high, and he may reoffend within three months. Another offender may be certain that he will return to prison very soon, but matters take another course and he stays out of prison, to his own surprise. So, using a neuroprediction technique in order to assess the risk of recidivism may not be an expression of distrust. The examiner tries to obtain information about something that is not known to the subject (even though he may have a hunch). Accordingly, the distinction knowledge/ignorance, understood in this way, is relevant to the issue of *trust*.

But it is also relevant in another way. Brain-reading may, in future, be used in different areas of society for different purposes. In criminal law, it may well be used in exactly those circumstances where the subject is *not trusted* (since she is believed to be hiding something) or is reluctant to share information. It is likely that in such circumstances, subjects (e.g. defendants) may also be *reluctant* to undergo brain-reading. In other words, they may not be cooperative. Therefore, in the context of criminal law and forensic psychiatry, it is very relevant to what extent these technologies could be used with some form of coercion (see below).

The second distinction concerns whether *cooperation* of the subject with the application of the test is required, and if so, to what extent. Some brain-reading technologies clearly require some form of collaboration: lie-detection is only possible as long as the subject is willing and ready to make statements. If the subject would remain silent, there are no utterances regarding which the honesty or dishonesty can be established. Note that such a type of brain-reading cannot be used if the subject completely refuses cooperation. On the other hand, and this is ethically and legally salient, the subject could, at least in principle, be coerced, or the examiner could try to make the subject cooperate. It might even be possible that in the future, brain-reading devices become available that do not require a person's cooperation. It might be possible to—unbeknownst to the subject—detect aspects of the mind of that person—we leave open to the reader's imagination how this might become possible. Another possibility is that a person is sedated and undergoes a type of brain scanning that does not require the subject to be aware and perform a task (just like a brain MRI may be made in an unconscious patient).

Clearly, more distinctions could be made—and the distinctions we have made could be elaborated upon. But within the context of this chapter—which aims to identify some areas of ethical and, in particular, legal interest—we will limit ourselves to these central distinctions.

Still, there will be one additional aspect of brain-reading that is of interest: the *nature and amount* of information that is obtained. For instance, in P300 research employed with EEG, the information may be no more than the recognition of a picture. This may not be very privacy-sensitive. Marcel Just and his colleagues were able to identify—to some extent—the physics concepts (such as acceleration, temperature) subjects were thinking about in the scanner (Mason & Just, 2016). This does not seem too privacy-sensitive either. However, this line of research also focusses on the neural representation of emotions, political orientation and suicidal ideation (Kassam et al., 2013; Leshinskaya et al., 2017; Just et al., 2017). Obviously, this is privacy-sensitive information.

Central Issues in the Ethical and Legal Debate: Learning from Each Other

In this section, we briefly outline three *ethical* concerns regarding brain-reading in forensic psychiatry and show whether and, if so, how similar issues are discussed in the *legal* literature. In doing so, we illustrate how both ethics and the law can learn from one another, providing helpful insights to push forward the legal-ethical debate on brain-reading, justice and security. The three ethical issues central to this section relate to (1) autonomy, (2) confidentiality and (3) trust (see also Meynen, 2017, 2019). Note that this is just an indication of ethical issues related to brain-reading in forensic psychiatry—more can be said about each of them.

Autonomy

From a medical-ethical perspective, regarding any medical procedure, the patient's autonomous informed consent is crucial. Yet, sometimes, non-consensual interventions may be imposed, such as quarantine in some cases of infectious diseases. Also in psychiatry, sometimes people are involuntarily admitted to a mental hospital, or they may be sentenced by a criminal court to forensic psychiatric care. In other cases, people may even be forced to undergo certain treatment, such as antipsychotic treatment. Perhaps, in the near future, the use of brain-reading may be considered as an option in forensic psychiatry. Could a forensic patient or a defendant be physically compelled to undergo such brain-reading? This will not only depend on ethical considerations, but clearly also on the type of technology: if some level of cooperation is required, the procedure cannot be physically enforced. If cooperation is not required, however, it could be technically possible to make that person undergo the brain-reading. But, even if the procedure cannot be enforced upon a person, we may coerce a person to cooperate with the brain-reading in psychiatry, by threatening with negative consequences for those who refuse to cooperate. As Szmukler and Appelbaum make clear, in psychiatry, there is a range of '(semi)coercive' options—leverage, coercion, and compulsion—that may make a patient 'willing' to cooperate (Szmukler & Appelbaum 2008). What level of pressure would be ethically permissible for different types of brain-reading—if any?

In the legal debate, autonomy is an important notion as well.[6] For example, the permissibility of *non-consensual* forensic brain-reading is

[6]In our legal considerations, we focus on the ECHR.

under debate in the light of the right to privacy and data protection, as well as regarding the right to remain silent during a criminal trial.[7] According to the European Court on Human Rights (ECtHR/the Court), personal autonomy is an important consideration underlying the interpretation of the right to privacy pursuant to Article 8 of the European Convention on Human Rights (ECHR).[8] In addition, the right to silence and freedom from self-incrimination partly follow from the autonomy of the individual in adopting a particular defence strategy (Harris et al., 2018, p. 423). However, whereas in ethics, the principle of autonomy may provide concrete guidance for regulating new technologies such as brain-reading, e.g. by demanding particular requirements for valid consent (Brownsword, 2012; Edwards, 2012; Lavazza, 2018), in law, autonomy has a more abstract role, e.g. as a background consideration or a relevant factor in interpreting legal rights. Moreover, restricting autonomy is not uncommon in criminal law, for example, by imposing criminal sentences or obliging a witness to testify truthfully in court. It is relevant to note that the legal definition of autonomy can be different across legal systems, and it does not necessarily correspond to the concepts of autonomy in medical ethics or moral philosophy. For example, the ECtHR defines the right to autonomy as 'the right to make choices as to how to lead one's own life',[9] which has been at stake in cases on abortion, voluntary euthanasia and recognition of transsexuals.[10] It is important to realize that such a legal interpretation is not necessarily identical to the understanding of autonomy in, e.g. the Kantian sense. Hence, whereas from an ethical perspective, an argument from autonomy might provide a valid claim for, or against non-consensual brain-reading, we should be careful in extrapolating the same argument to the legal debate, realising that 'autonomy' can mean different things in different contexts.

---

[7]See e.g. Ligthart (2019) and Shen (2013).

[8]ECtHR (GC) 5 September 2017, appl.no. 61496/08 (*Bărbulescu/Romania*), § 70; ECtHR (GC) 15 March 2012, appl.nos. 4149/04 and 41029/04 (*Aksu/Turkey*), § 58.

[9]ECtHR 3 September 2015, appl.no. 10161/13 (*M. and M. v. Croatia*), § 171.

[10]ECtHR (GC) 16 December 2010, appl.no. 25579/05 (*A, B and C v. Ireland*), § 216; ECtHR 29 April 2002, appl.no. 2346/02 (*Pretty v. UK*), § 61–67; ECtHR 11 July 2002, appl.no. 25680/94 (*I./UK*), § 70–73.

Although the law should be careful in drawing upon ethical claims, ethics could provide helpful insights in the development of a legal right to mental autonomy. One particular area of the law where such an integrative approach has been employed and might be further developed concerns the notion of informed consent (e.g. Beyleveld & Brownsword, 2007; Buelens et al., 2016). For example, in general, European human rights such as the right to privacy and the prohibition of ill-treatment (Articles 8 and 3 ECHR) do normally not protect from state interference if the individual consented with the interference at issue (Buelens et al., 2016). A medical intervention, for instance, that seriously interferes with the subject's bodily integrity will not raise an issue under the prohibition of ill-treatment if the subject gave his informed consent to it.[11] As was briefly mentioned above, forensic brain-reading can be deployed using different types of pressure, ranging from making serious threats (e.g. if you do not cooperate with this brain-reading test, your detention will be extended) to offering just another option (e.g. if you cooperate with this brain-reading test, you have to report to the parole officer less frequently). To what extent do such threats and offers respect the voluntariness of the individual's consent for methods of criminal investigation, and how should the law approach situations like these? As yet, at the level of European human rights, no in-depth legal theoretical approach exists on this particular issue (Ligthart, 2020a, pp. 160–165). By contrast, medical ethics and moral philosophy provide lengthy and in-depth debates on threats, (coercive) offers and the validity of consent (e.g. Kiener, 2020; Eyal, 2019; Anderson, 2017), also with respect to neurotechnology (e.g. Brownsword, 2012; Edwards, 2012; Pugh, 2018). Hence, in developing a legal approach on the voluntariness and validity of consent in the context of deploying forensic brain-reading, the theoretical ethical and philosophical debate might be inspiring for the law on this particular issue.

[11]ECtHR 3 October 2008, appl.no. 35228/03 (*Bogumil/Portugal*), § 71.

Confidentiality

There is something special about forensic assessments of defendants. Whereas in standard medical practice there is strict confidentiality regarding the information the physician obtains about the patient (even though there are limitations), a forensic psychiatric report will be shared with the court (if it is being made for the court). Moreover, the information may not only be shared with the court, the prosecutor and the defence attorney, but it may also become known to journalists who attend the courtroom proceedings while the psychiatrist gives testimony (Meynen, 2019). Present and future brain-reading applications may provide information about all kinds of mental phenomena of a person, some very intimate. Which information should be shared with the court? It may be that certain regulations already cover sharing of information in certain jurisdictions in a way that can be used in these situations. But it is good to realize that currently, forensic psychiatric evaluations have a question–answer structure (Meynen, 2019). In practice, this structure limits the range of topics and thus answers by the examinee. Brain-reading could more broadly pick up mental contents that is not 'an answer to a question' but which just happens to be on a person's mind. In addition, the data yielded through *prima facie* question–answer structured brain-reading, such as lie-detection, might be (re-)analysed and interpreted in different ways that enable drawing inferences about other mental states, e.g. for purposes of neuroprediction. Depending on the nature and scale of the eventually acquired data, this typical feature of brain-reading—that it may yield unexpected and very private information—may create new challenges for sharing information. In addition, it is clear that obtaining—potentially large-scale—highly confidential information through brain scanning technology about a person requires that this will be dealt with according to the regulations for data storage.

In law, the notion of confidentiality translates itself into different human rights. First, the right to respect for informational privacy pursuant to Article 8 ECHR protects the individual's personal data. In general, the level of legal protection in this context depends on the amount of information at stake and its level of sensitivity. For example, the information yielded with a P300-test employed with EEG, enabling to reveal whether the person recognizes a particular object, such as a gun, might be considered less sensitive than the results of an MRI-scan, disclosing that one suffers from brain cancer (Ligthart, 2019). Another human right that echoes the protection of 'confidentiality' of certain mental states is the right to freedom of thought as guaranteed by Article 9 ECHR (Bublitz, 2014; McCarthy-Jones, 2019, Ligthart, 2020b). Apart from state interferences aiming to *control* a person's thoughts, convictions and religion, Article 9 ECHR also prohibits coercive measures to *disclose* thoughts, conscience or religion (e.g. Vermeulen & Roosmalen, 2018, p. 738; Harris et al., 2018, pp. 574–575; Taylor, 2005, p. 120).[12] However, since the Grand Chamber of the ECtHR held that 'the right to freedom of thought, conscience and religion denotes only those views that attain a certain level of cogency, seriousness, cohesion and importance',[13] it is not clear at all that the results of forensic brain-reading, such as whether one indeed hears commanding voices (lie-detection) or has a 'high risk' brain feature (neuroprediction) qualifies as a thought in the meaning of Article 9 ECHR (Ligthart et al., 2020). For example, as Evans notes, the right to freedom of thought and conscience embraces personal thoughts on political, philosophical, ethical and intellectual positions in human affairs (Evans 2001, p. 52). Similarly, Partsch argues that freedom of thought concerns 'political and social thought' (Partsch, 1981, pp. 213–214). Whereas the results of a forensic brain-reading test can indeed be very important for the individual at issue, suffering from a particular brain feature and lying about commanding voices can hardly be considered as 'views that attain a certain level of cogency, seriousness, cohesion and importance', similar to, for example, philosophical and political attitudes. Hence, as in the context of the right to privacy pursuant to Article 8 ECHR, the legal protection that the right to freedom of thought offers regarding forensic brain-reading, probably also depends on the *content* of the acquired results—that is, whether the yielded information enables the drawing of inferences about one or more

[12]See e.g. ECtHR 3 June 2010, appl.nos. 42837/06, 3237/07, 3269/07, 35793/07 and 6099/08 (*Dimitras and others/Greece*); ECtHR 2 May 2010, appl.no. 21924/05 (*Sinan Işık/Turkey*).

[13]ECtHR (GC) 1 July 2014, appl.no. 43835/11 (*S.A.S./France*), § 55. See also ECtHR (GC) 26 April 2016, appl.no. 62649/10 (*İzzettin Doğan and others v. Turkey*), § 68.

personal views that, given their content, attain a certain level of cogency, seriousness, cohesion and importance (Ligthart, 2020b).

From an ethical perspective, it has been argued that the current framework of European human rights does not adequately protect the confidentiality of peoples' mental states that can be disclosed with the use of emerging brain-reading technologies. To amend this situation, some have proposed the recognition of a novel human right to 'mental privacy', which should basically protect any bit of brain data that can be acquired through current and futuristic brain-reading technologies (Ienca & Andorno, 2017; Lavazza, 2018).

Indeed, the existing human rights that are (partly) created to protect the confidentiality of personal information, i.e. the right to privacy and freedom of thought, might not prohibit all forensic brain-reading applications. Yet, other existing fundamental rights, that are not typically created to protect the notion of *confidentiality* but, for instance, rather follow from the concept of autonomy, might protect from non-consensual brain-reading as well, potentially filling the supposed 'gaps' in human rights protection. More specifically, as to defendants in criminal trials, being coerced to participate in a forensic brain-reading test, i.e. revealing brain data that contributes to one's own conviction and/or sentencing, raises issues under the privilege against self-incrimination as guaranteed by Article 6 ECHR (Meijer & Van Toor, 2021, in this volume).

In addition, the right to freedom of expression pursuant to Article 10 ECHR seems to imply a negative right for each individual not to convey information, ideas and opinions (Harris et al. 2018, p. 595). For example, in different cases, the European Commission of Human Rights held that "the right to freedom of expression by implication also guarantees a 'negative right' not to be compelled to express oneself, i.e. to remain silent".[14] As yet, the Grand Chamber has not explicitly acknowledged such a right to non-expression under Article 10 ECHR, but 'does not rule out that a negative right to freedom of expression

[14]E.g. EComHR 7 April 1994, appl.no. 20871/92 (*Strohal/Austria*); EComHR 1 March 1993, appl.no. 17488/90 (*Goodwin/UK*); EComHR 13 October 1992 appl.no. 16002/90 (*K./Austria*).

is protected under Article 10 of the Convention'.[15] Since the scope of Article 10 ECHR is considered 'very broad' (Rainey et al., 2017, p. 483), encompassing information of almost any content, conveyed through any means (Lester, 1993, p. 469; Harris et al., 2018, p. 594), the right to freedom of (non-)expression could potentially also cover the conveyance of information through brain-reading technologies.

Hence, before the law should create a novel human right to mental privacy, protecting the confidentiality of brain data as argued by some ethical scholars, the precise legal implications of existing rights should be closely examined, including the privilege against self-incrimination pursuant to Article 6 ECHR, and the right not to convey information, ideas and opinions under Article 10 ECHR. This example illustrates that whereas ethics can indeed be informative for the law, legal doctrines can also inform ethics, e.g. in ethical claims of developing novel human rights.

Trust

In forensic psychiatry, patients are not only evaluated, but many are treated as well. Trust is an important—if not crucial—element of a treating relationship. Taking into account the relevance of trust in the above analysis (section "The Relevance of Brain-Reading for Forensic Psychiatry and Criminal Proceedings"), the use of brain-reading technologies that try to obtain information that the patient already knows is, as said, likely to convey the message: 'I don't trust you' or at least it can be understood in this way. This may undermine a trusting relationship in a sometimes long and intensive treatment trajectory in forensic psychiatry. Still, one could argue that, in forensic psychiatry, there will often be a natural attitude in which one is wary of the possibility that a person may be untruthful. Grubin has pointed to positive effects of (classical polygraph) lie-detection in forensic psychiatry:

> For the forensic patient, polygraphy offers the opportunity to demonstrate that he is low risk, and it can encourage him to cooperate with

---

[15] ECtHR (GC) 3 April 2012, appl.no. 41723/06 (*Gillberg/Sweden*), § 86.

> treatment and management plans by making it explicit when he is not. It also allows intervention to prevent an increase in risk or relapse in symptoms. Although some may be worried that it will affect the therapeutic relationship with the patient, there is no evidence to suggest such an effect. After all, the aim is to encourage truth-telling rather than to catch the patient out in a lie. (Grubin, 2010, p. 450)

So, in a way, lie-detection might actually provide a way for a patient to show his trustworthiness (see also Meynen, 2017). Still, it is good to realize that, as far as the issue of trust is concerned, the relationship between the forensic psychiatrist and forensic patient is not the same as a policeman–suspect relationship.

In criminal law, the ethical notion of trust cannot be directly translated into an explicit 'right of trust' to the benefit of the accused person (or, at the trial stage, the defendant). Nevertheless, like all citizens within jurisdictions governed by the Rule of Law, a suspected person should be able to count on the fact that the authorities, when applying investigative methods, will not exceed the limits that have been set by the law. In this regard, the principle of legality is an important safeguard against arbitrary actions by the authorities (Corstens/Borgers & Kooijmans, 2018, p. 23). Due to this important principle, investigative methods that restrict human rights and freedoms should be described in a law which is sufficiently accessible and foreseeable.[16] The democratic aspect of the principle of legality should, in addition, guarantee that citizens are not subjected to disproportionate investigative methods. Citizens, including defendants, should be able to trust the government in this respect. As a consequence of the principle of legality, investigation methods in criminal procedure—including the use of brain-reading technologies—should be accurately described by law, in order to enable the citizens to know in which circumstances and under which conditions those methods can be applied.

As mentioned before, the relationship between the forensic psychiatrist and forensic patient is not the same as a policeman-suspect relationship. The same can be said about the relationship between the accused

[16]See *inter alia* ECtHR 26 April 1979, appl.no. 6538/74 (*Sunday Times/UK*).

person and the probation and after–care service. In several countries, like the Netherlands, officials of this institution (*Reclassering*) already visit the accused person during the early stage of police custody in order to compose a preliminary report. The aim of this report is to provide the authorities of the public prosecution service and the judiciary with information about the suspect. If the contents of this report would be used as evidence in the criminal case, the (future) relationship between *Reclassering*-officials and suspects would be jeopardized, which could affect the usefulness of those reports. Therefore, the report is excluded from evidence in the Netherlands.[17] The suspect can trust that nothing he says during the interview with the *Reclassering*-official will be used against him as far as the proof of the allegation is concerned.

One other aspect of trust in the legal domain should be mentioned briefly. In theory, apart from defendants, brain-reading technologies could also be deployed in order to verify the accuracy of statements of *victims* who report a crime to the police. In fact, in Japan, memory detection can be employed if the investigator suspects that the victim's complaint is false (Osugi, 2011).[18] In general, the application of those technologies could also convey the message: 'I don't trust you' as a consequence of which the willingness to report crimes could be undermined and, therefore, crimes could remain unresolved. This could harm the general, societal trust in criminal law and the effectiveness of criminal proceedings.

Conclusion

In this chapter, we explored some commonalities in the ethical and legal debates on brain-reading in forensic psychiatry and criminal justice. We briefly identified three central issues from the ethical debate and explored whether and how these issues are reflected in the legal discussion as well.

[17]Supreme Court of the Netherlands 18 September 2007, ECLI:NL:HR:2007:BA3610. Cf. Supreme Court of the Netherlands 25 September 2012, ECLI:NL:HR:2012:BX4269.

[18]Notice that in Japan the concealed information test is not applied with a brain-reading technique, but with a polygraph, measuring physiological reactions of the autonomous nervous system.

We illustrated that in ethics and the law similar (interrelated) issues arise regarding brain-reading technologies aiming to disclose particular mental states—that is, issues of autonomy, confidentiality and trust. However, as we noted, although these central themes are reflected in both the ethical and legal debate, the way in which ethics and law provide a normative framework for the application brain-reading in the context of criminal justice may vary. Although ethical claims can be informative for the law, for example, in developing the right to personal autonomy, we should be careful in extrapolating ethical arguments into the legal debate. Reversely, legal doctrines can—and should—sometimes inform ethics as well. For example, as illustrated, ethical claims for a novel human right to mental privacy should take into account legal research on the level of legal protection that the current framework of human rights offers in this respect. By critically informing each other's disciplines in the debate on forensic brain-reading, the ethical and legal debate could strengthen each other, pushing forward research on neuroscience, justice and security.

References

Aharoni, E., et al. (2013). Neuroprediction of future rearrest. *PNAS, 110*(15), 6223–6228.

Alimardani, A., & Chin, J. (2019). Neurolaw in Australia: The use of neuroscience in Australian criminal proceedings. *Neuroethics, 12*(3), 255–270.

Anderson, S. (2017). Coercion. In E. N. Zalta (Ed.), *The Stanford Encyclopedia of Philosophy* (Winter 2017 ed.).

Beyleveld, D., & Brownsword, R. (2007). *Consent in the law*. Hart Publishing.

Brownsword, R. (2012). Regulating brain imaging: Questions of privacy, informed consent, and human dignity. In S. Richmond, G. Rees, & S. J. L. Edwards (Eds.), *I know what you're thinking*. Oxford University Press.

Bublitz, J. C. (2014). Freedom of thought in the age of neuroscience. *Archiv Für Rechts- Und Sozialphilosophie, 100,* 1–25.

Buelens, W., Herijgers, C., & Illegems, S. (2016). The view of the European Court of Human Rights on competent patients' right of informed consent.

Research in the light of Article 3 and 8 of the European Convention on Human Rights. *European Journal of Health Law, 23*(5), 481–509.

Catley, P., & Claydon, L. (2015). The use of neuroscientific evidence in the courtroom by those accused of criminal offenses in England and Wales. *Journal of Law and the Biosciences, 2*(3), 510–549.

Corstens, G. J. M. (2018). *Het Nederlands strafprocesrecht*, negende druk, bewerkt door M. J. Borgers en T. Kooijmans. Wolters Kluwer.

Cruft, R., Liao, M., & Renzo, M. (Eds.). (2015). *The philosophical foundations of human rights*. Oxford University Press.

De Kogel, C. H., & Westgeest, E. J. M. C. (2015). Neuroscientific and behavioral genetic information in criminal cases in the Netherlands. *Journal of Law and the Biosciences, 2*(3), 580–605.

Delfin, C., et al. (2019). Prediction of recidivism in a long-term follow-up of forensic psychiatric patients: Incremental effects of neuroimaging data. *PLoS ONE, 14*(5), e0217127.

Edwards, S. J. L. (2012). Protecting privacy interests in the brain images: The limits of consent. In S. Richmond, G. Rees, & S. J. L. Edwards (Eds.), *I know what you're thinking*. Oxford University Press.

Evans, C. (2001). *Freedom of religion under the European Convention on Human Rights*. Oxford University Press.

Eyal, N. (2019). Informed consent. In E. N. Zalta (Ed.), *The Stanford Encyclopedia of Philosophy* (Spring 2019 ed.).

Farah, M. J., et al. (2014). Functional MRI-based lie detection: Scientific and societal challenges. *Nature Reviews Neuroscience, 14*, 123–131.

Farahany, N. A. (2015). Neuroscience and behavioral genetics in US criminal law: An empirical analysis. *Journal of Law and the Biosciences, 2*(3), 485–509.

Grubin, D. (2010). The polygraph and forensic psychiatry. *Journal of the American Academy of Psychiatry and the Law, 38*, 446–451.

Hafner, M. (2019). Judging homicide defendants by their brains: An empirical study on the use of neuroscience in homicide trials in Slovenia. *Journal of Law and the Biosciences, 6*(1), 226–254.

Harris, D. J., et al. (2018). *Harris, O'Boyle, and Warbrick: Law of the European Convention on Human Rights*. Oxford University Press.

Ienca, M., & Andorno, R. (2017). Towards new human rights in the age of neuroscience and neurotechnology. *Life Sciences, Society and Policy, 13*(5), 1–27.

Just, M. A., et al. (2017). Machine learning of representations of suicide and emotion concepts identifies suicidal youth. *Nature Human Behavior, 1,* 911–919.

Kassam, K. S., et al. (2013). Identifying emotions on the basis of neural activation. *PLoS ONE, 8*(6), e66032.

Kiener, M. (2020). Coercion. In E. Craig (ed.), *Routledge Encyclopedia of Philosophy* (Version 2, 2020).

Lavazza, A. (2018). Freedom of thought and mental integrity: The moral requirements for any neural prosthesis. *Front Neurosci, 12*(82).

Leshinskaya, A., et al. (2017). Neural representations of belief concepts: A representational similarity approach to social semantics. *Cerebral Cortex, 27*(1), 344–357.

Lester, A. (1993). Freedom of expression. In R. St. J. Macdonald et al. (Eds.), *The European system for the protection of human rights.* Martinus Nijhof Publishers.

Ligthart, S. (2019). Coercive neuroimaging, criminal law and privacy: A European perspective. *Journal of Law and the Biosciences, 6*(1), 289–309.

Ligthart, S. (2020a). Coercive forensic neuroimaging and the prohibition of ill-treatment (article 3 ECHR). In A. Waltermann et al. (Eds.), *Law, science and rationality.* Eleven Publishers.

Ligthart, S (2020b). Freedom of thought in Europe: Do advances in brain-reading technology call for revision? *Journal of Law and the Biosciences,* lsaa048.

Ligthart, S., Douglas, T., Bublitz, J. C., Kooijmans, T., & Meynen, G. (2020). Forensic brain-reading and mental privacy in European human rights law: Foundations and challenge. *Neuroethics* (online first).

Ligthart, S., Douglas, T., Bublitz, J. C., & Meynen, G. (2019). The future of neuroethics and the relevance of the law. *AJOB Neuroscience, 10*(3), 120–121.

Mason, R. A., & Just, M. A. (2016). Neural representation of physics concepts. *Psychological Science, 27*(6), 904–913.

McCarthy-Jones, S. (2019). The autonomous mind: The right to freedom of thought in the twenty-first century. *Frontiers in Artificial Intelligence, 2*(19), 1–17.

Meijer, E. H., & Van Toor, D .A. G. (2021). Reading the sleeping mind: Empirical and legal considerations. In D. A. G. Van Toor et al. (Eds.), *Neurolaw: Ways forward for neuroscience, justice, and security.* Palgrave Macmillan.

Meijer, E. H., et al. (2016). Deception detection with behavioral, autonomic, and neural measures: Conceptual and methodological considerations that warrant modesty. *Psychophysiology, 53,* 593–604.

Meynen, G. (2017). Brain-based mind reading in forensic psychiatry: Exploring possibilities and perils. *Journal of Law and the Biosciences, 4*(2), 311–329.

Meynen, G. (2018a). Forensic psychiatry and neurolaw: Description, developments, and debates. *International Journal of Law and Psychiatry, 65,* 101345.

Meynen, G. (2018b). Author's response to peer commentaries: Brain-based mind reading: Conceptual clarifications and legal applications. *Journal of Law and the Biosciences, 5*(1), 212–216.

Meynen, G. (2019). Ethical issues to consider before introducing neurotechnological thought apprehension in psychiatry. *AJOB Neuroscience, 10*(1), 5–14.

Meynen, G. (2020). Neuroscience-based psychiatric assessments of criminal responsibility: Beyond self-report? *Cambridge Quarterly of Healthcare Ethics, 29,* 446–458.

Morse, S. J., & Roskies, A. L. (Eds.). (2013). *A primer on criminal law and neuroscience*. New York: Oxford University Press.

Osugi, A (2011). Daily application of the concealed information test: Japan. In B. Verschuere, G. Ben Shakhar, & E. Meijer (Eds.), *Memory detection: Theory and application of the concealed information test*. Cambridge University Press.

Pardo, S., & Patterson, D. (2015). *Minds, brains, and law. The conceptual foundations of law and neuroscience.* Oxford University Press

Partsch, K. J. (1981). Freedom of conscience and expression, and political freedoms. In L. Henkin (Ed.), *The International Bill of Rights: The covenant on civil and political rights.* Columbia University Press.

Pugh, J. (2018). Coercion and the neurocorrective offer. In D. Birks & T. Douglas (Eds.), *Treatment for crime: Philosophical essays on neurointerventions in criminal justice*. Oxford University Press.

Rainey, B., Wicks, E., & Ovey, C. (2017). *The European Convention on Human Rights.* Oxford University Press.

Richmond, S., Rees, G., & Edwards, S. J. L. (Eds.). (2012). *I know what you're thinking*. Oxford University Press.

Shen, F. X. (2013). Neuroscience, mental privacy and the law. *Harvard Journal of Law & Public Policy, 36,* 653–713.

Simpson, J. R. (Ed.). (2012). *Neuroimaging in forensic psychiatry: From the clinic to the courtroom.* Wiley-Blackwell.

Szmukler, G., & Appelbaum, P. S. (2008). Treatment pressures, leverage, coercion, and compulsion in mental health care. *Journal of Mental Health, 17*(3), 233–244.

Taylor, P. M. (2005). *Freedom of religion: UN and European human rights law and practice*. Cambridge University Press.

Vermeulen, B., & Roosmalen, M. (2018). Freedom of thought, conscience and religion. In P. Van Dijk et al. (eds.), *Theory and practice of the European Convention on Human Rights*. Intersentia.

Sjors Ligthart holds a Master's in Criminal Law. He is currently completing his PhD thesis on coercive brain-reading in criminal law. Other research interests include the introduction of neurointerventions and virtual reality systems into the domain of criminal justice, the debate on fundamental neurorights, and the concept of legal insanity. He is a lecturer of Penitentiary Law and editorial secretary of the leading Dutch journal on criminal law.

Tijs Kooijmans is a full Professor of Criminal Law at Tilburg University and a substitute judge at the 's-Hertogenbosch Court of Appeals. He is a co-author of a leading handbook about Dutch criminal procedure and a commentator of Dutch leading criminal cases. His research interest lies in criminal law in general, confiscation and seizure of illegally obtained assets, forensic psychiatry, and neurolaw.

Gerben Meynen is Psychiatrist and Professor of Forensic Psychiatry, Willem Pompe Institute for Criminal Law and Criminology (Utrecht Centre for Accountability and Liability Law, UCALL), Utrecht University, and professor of Ethics and Psychiatry, Department of Philosophy, Humanities, Vrije Universiteit Amsterdam. His research interests include legal insanity and the implications of neuroscience for criminal law and forensic psychiatry.

Ethical Perspectives

A Biopsychosocial Approach to Idiopathic Versus Acquired Paedophilia: What Do We Know and How Do We Proceed Legally and Ethically?

Cristina Scarpazza, Colleen Berryessa, and Farah Focquaert

Introduction

On June 2016, a 30-year-old man (R.H.) was given 22 life sentences for serious sexual assault against a minimum of 71 children. During the hearings, R.H. revealed some of the stratagems he employed to procure victims, such as taking children out on day trip from foster homes and

C. Scarpazza (✉)
Department of General Psychology, University of Padova, Padova, Italy
e-mail: cristina.scarpazza@unipd.it

C. Berryessa
School of Criminal Justice, Rutgers University, Newark, NJ, USA
e-mail: colleen.berryessa@rutgers.edu

F. Focquaert
Department of Philosophy and Moral Sciences, Ghent University, Ghent, Belgium
e-mail: farah.focquaert@ugent.be

S. Ligthart et al. (eds.), *Neurolaw*, Palgrave Studies in Law, Neuroscience, and Human Behavior,
https://doi.org/10.1007/978-3-030-69277-3_7

escorting them home from their own birthday party. He produced child pornography, and he shared photos of his crimes with other paedophiles. He was proven to be fully aware of his behaviours that were carried out in a logical, reasoned and predatory way. He said that "impoverished kids are definitely much easier to seduce than middle-class kids", revealing his thoughtful selection of his victims. His neurological and neuropsychological examinations were normal.

In 2011, a 64-year-old paediatrician without previous criminal record was charged with child abuse (Sartori et al., 2016; et al., 2018b) as he was found while enacting sexually inappropriate behaviour towards a child in a kindergarten doctor's office, leaving the door open. Upon arrest, he was not aware of the social and legal implication of his action: he told his receptionist to postpone all his patients to the day after. He did not actively search for children, but his job put him in close contact with them. On neurological and psychiatric examinations, conducted while he was at house arrest, he showed asymmetrical brisk motor reflexes, along with symptoms and signs suggestive of optic chiasm compression (tunnel vision and diplopia) and frontal lobe dysfunction, including pathological crying, behaviorial dis-inhibition, easy irritability, childish and obsessive–compulsive behaviours and impairments in emotion attribution, moral reasoning and abstract thinking. For instance, while travelling with his wife, the patient would steal postcards from exhibitors in museum shops.

Paedophilia is a disorder of high public concern due to its association with child sexual offence and recidivism (Hall & Hall, 2007; Hanson, 2002; Hanson & Morton-Bourgon, 2005; Hanson et al., 1993, 2003; Seto, 2009; Seto et al., 2004). Although paedophilia is considered a relatively rare phenomenon (with a prevalence of 3–5% in the male population [Beech et al., 2016]), offenders with paedophilia commit a disproportionate amount of crime. Despite the growing literature on this disorder, it is still not widely known that paedophilia is not a unitary phenomenon. One of the less investigated categorizations within offenders with paedophilia refers to the difference between offenders with idiopathic and acquired paedophilic behaviour.

The first case description presented in our introduction refers to an offender with idiopathic paedophilic disorder, a psychiatric disorder included within the section of paraphilias in the DSM-5. In the case description presented above, the paedophilic behaviour is the sole disorder manifested by the offender, making this a clear case of idiopathic paedophilia. Although idiopathic paedophilia is widely known and described in the literature (see, e.g., Hall & Hall, 2007; Seto, 2009; Tenbergen et al., 2015 for reviews), little is still known regarding acquired paedophilia (Camperio Ciani et al., 2019; Gilbert & Focquaert, 2015).

The term acquired paedophilia refers to the insurgence of paedophilic interest and behaviours in previously non-paedophilic men after a brain insult. Indeed, despite it being widely known that neurological disorders are commonly associated with psychiatric symptoms, it is less evident and clear that a number of neurological disorders can show a predominant behaviorial and sometimes bizarre presentation and for this reason can be mistakenly diagnosed as psychiatric (Butler & Zeman, 2005; Keshavan & Kaneko, 2013). This is the case with regard to acquired paedophilic behaviour. In such cases, paedophilia is one of the symptoms of an underlying neurological condition, as acquired paedophilia is always associated with additional cognitive, neurologic and behaviorial symptoms that depend on the underlying brain pathology. The second case description presented in our introduction refers to an offender with acquired paedophilia. Indeed, a magnetic resonance imaging examination revealed the presence of a clivus chordoma (a slow growing tumour of the notochord) that displaced the pituitary gland and compressed the orbitofrontal cortex, the optic chiasm and the hypothalamus. For this reason, the position of the tumour alone explains all the cognitive and behaviorial symptoms manifested by the patient.

Pertaining comprehensive knowledge of the differences between idiopathic and acquired paedophilia is of utmost importance from the medical, ethical and legal point of view. In the following paragraphs, a short summary of the available literature is presented to clarify to the readers the differences between these two forms of paedophilia. Indeed, acquired paedophilic behaviour differs from idiopathic paedophilic disorder in many aspects: aetiology, underlying neural correlates, *modus operandi*, possible therapies, legal implications for punishment. Although both disorders are in many ways different, there can be some similarities as well depending on the case at hand, as well as reasons to consider treatment rather than retributive punishment in both kinds of pathology.

Aetiology

The first important difference between idiopathic and acquired paedophilia lies in aetiology: while idiopathic paedophilia is categorized within psychiatric disorders, acquired paedophilia clearly has a neurological origin.

Idiopathic paedophilic disorder is considered to be a psychiatric disorder included within the paraphilias in the DSM5 (Beech et al., 2016). In DSM-5, paedophilia is de-pathologized as the manual underlines that paedophilia becomes a disorder when the sexual attraction towards children is paired with a significant distress and impairment by fantasies and urges, or the acting out on behaviorial level. In idiopathic paedophilia, the paedophilic interest would appear to be stable across the individual's lifespan (Hanson et al., 1993) and it typically first appears in adolescence (Tenbergen et al., 2015). As for all psychiatric conditions, paedophilia does not have a clear aetiology, and different psychological and environmental theories have been proposed (Doshi et al., 2018; Tenbergen et al., 2015). In particular, research regarding the aetiology of paedophilia suggests the presence of a complex and multifactorial phenomenon in which the influences of genetics (Kruger et al., 2019), stressful life events (Jespersen et al., 2009), testosterone exposure, neurochemical impairment (mainly serotoninergic disturbances) (Gilbert & Focquaert, 2015) as well as subtle brain alterations, may generate this specific phenotype of sexual preference (Cantor et al., 2008; Schiffer et al., 2007; Schiltz et al., 2007; Tenbergen et al., 2015). Early theories also considered the influence of psychological mechanisms such as the "abused-abuser" theory (Freund & Kuban, 1994; Freund et al., 1990; Hall & Hall, 2007) on the sexual preference of individuals with paedophilia. Indeed, the numbers reported for individuals with paedophilia who were abused as children range from 28 to 93% vs 15% for random controls (Cohen & Galynker, 2002; Greenberg et al., 1993; Hall & Hall, 2007). Furthermore, literature suggests that child sexual offending is characterized by emotional disturbances and a high rate of psychopathology (Kruger & Schiffer, 2011; Tenbergen et al., 2015), a high rate of social anxiety, less social engagement, low self-esteem and a decreased ability to be socially assertive (Geer et al., 2000; Hall &

Hall, 2007). Frequently idiopathic paedophilia has comorbidities with psychiatric disorders: for instance, 60% of individuals with paedophilia also qualified for a personality disorder (Fagan et al., 2002; Green, 2002; Kruger & Schiffer, 2011; Raymond et al., 1999).

To sum up, the aetiology of idiopathic paedophilic disorder is still unknown, but is considered to be multifactorial, with the influence of biological, psychological and social factors. In contrast, acquired paedophilic behaviour by definition refers to the insurgence of sexual urges towards children later in life as a consequence of an acquired neurological condition with a clear neurologic aetiology. Cases of paedophilia associated with brain damage have been described in patients with frontotemporal dementia (Mendez, 2010), brain tumours (Burns & Swerdlow, 2003), clivus chordoma (Sartori et al., 2016), surgical lesions (Devinsky et al., 2010), hippocampal sclerosis (Mendez & Shapira, 2011), multiple sclerosis (Frohman et al., 2002), etc. These neurological insults seems to produce a "*behavioral fracture*" in the overt behaviour manifested prior and after the brain disease insurgence (Scarpazza et al., 2018a, 2018c). To further discuss the causal role of neurological disorders on the insurgence of paedophilic behaviours, two cases are of particular relevance (Burns & Swerdlow, 2003; Sartori et al., 2016). In both cases, paedophilia emerged as a symptom of a tumour: a *clivus chordoma* (a slow growing tumour of the notochord, in this case displacing the hypothalamus and compressing the orbitofrontal cortex (Sartori et al., 2016) and an hemangiopericytoma in the right orbitofrontal cortex (Burns & Swerdlow, 2003). In both cases, a *restitutio and integrum* of the symptomatology, including paedophilic urges, was documented after the surgical resection of the tumour, decreeing the causal link between the brain tumour and the paedophilic urges. In both cases, the tumour regrowth was accompanied by a re-insurgence of paedophilic interest, and a second surgical resection resulted again in a disappearance of the symptoms. To summarize, a clear aetiology is always present in offenders with acquired paedophilia. This aetiology depends on the underlying neurological condition: for instance, if acquired paedophilic disorder emerges during the course of dementia, the aetiology is neurodegenerative, while if it emerges following a brain injury the aetiology is traumatic.

Neural Correlates

The second important difference between idiopathic and acquired paedophilic behaviour lies in their neural correlates. While idiopathic paedophilia is associated with inconsistent subtle structural and/or functional abnormalities, individuals with acquired paedophilia clearly show some evident neuroanatomical alteration that, despite being spatially heterogeneous, localizes to a single functional network.

In line with its psychiatric aetiology, idiopathic paedophilia is characterized by functional brain alterations or subtle structural alterations without evident neuroanatomical abnormalities (as for instance, brain tumours or lesions) (Mohnke et al., 2014). Indeed, psychiatric disorders have long been considered "functional" disorders, without a significant structural neural substrate. Despite the fact that in the last two decades neuroimaging research using sophisticated statistical analysis on neuroimaging data, revealed that it is possible to observe neuroanatomical abnormalities in psychiatric disorders as well, literature has so far failed to identify a clinically useful neuroanatomical substrate for most psychiatric disorders, whom are still devoid of reliable biomarkers. This is true for paedophilia as well: quantitative voxel-based morphometric studies demonstrated volume reductions of the right amygdala, the hypothalamus and septal regions (Poeppl et al., 2013; Schiltz et al., 2007), structural deficits of temporal cortices and fiber bundles (Cantor et al., 2008; Schiffer et al., 2007) and morphologic abnormalities of the orbitofrontal cortex and basal ganglia (Schiffer et al., 2007). Further alterations appeared in areas in the parietal lobe (Cantor et al., 2008; Schiffer et al., 2007) as well as the cingulate cortex, the insula and cerebellum (Schiffer et al., 2007) when comparing paedophilic with non-paedophilic men. These alterations seem to be congenital or to emerge very early during life, encompassing brain regions involved in sexual arousal (Tenbergen et al., 2015) such as the amygdala and the hypothalamus. A recurrent finding is the overlap in functional brain activity according to the neurophenomenological model of sexual arousal when comparing paedophilic men to non-paedophilic men. There is comparable activity of the sexual arousal network when these individuals are confronted with sexual stimuli. Typically, paedophilic men do not show activity in the sexual arousal network when seeing pictures of naked adults, whereas non-paedophilic men typically do no show activity in the sexual arousal network when seeing pictures of naked prepubescent children (for instance, see Ponseti et al., 2012). The functional and structural brain alteration in idiopathic paedophilia is summarized in two recent reviews (Mohnke et al., 2014; Tenbergen et al., 2015).

A recent study (Scarpazza et al., 2021) described the results of a coordinate based meta-analysis run on previous structural and functional results of idiopathic paedophilia and underscores the absence of consistent and convergent results reporting brain abnormalities in offenders with paedophilia. Critically, this denotes the absence of a reliable neural underpinning for idiopathic paedophilic disorders. This differentiates paedophilia from psychiatric diagnoses where consistent structural and/or functional brain alterations have been identified in meta-analyses (Goodkind et al., 2015; Sha et al., 2019), even if these correlates are not clearly visible by the naked eye.

Two critical points are worth noting: first, the absence of consistent results across the existing literature makes it very difficult to relate group to individual inferences. In other words, this inconsistency of results makes it very difficult to accept the idea that results obtained at the level of the group could have a pathological meaning at the level of the single individual. It is thus still unknown whether the results obtained at the level of the group might have translational clinical implications. Second, as paedophilia has a high comorbidity with other psychiatric disorders (Eher et al., 2019), it is still not possible to disentangle whether the inconsistent results obtained so far truly reflect subtle neuroanatomical alterations of idiopathic paedophilia (e.g. reduced amygdala volume) or whether these results are more likely to reflect structural alterations of comorbid psychiatric disorders.

However, as mentioned, the case of acquired paedophilia is very different, where clear structural brain alterations emerging later in life are necessary for the diagnosis. In acquired paedophilia, neuroanatomical alterations, both lesions or atrophy, are clearly evident in each individual patient, and inferences can therefore be made for each patient. Crucially, these alterations have a causal link with the insurgence of paedophilic urges as there is a temporal link between the insurgence of brain alteration and the insurgence of paedophilic behaviour. Moreover, some rare cases have indicated that removing the brain alteration, for instance the tumour, led to a *restitution ad integrum* that included the disappearance of the paedophilic urges and behaviours (Burns & Swerdlow, 2003; Sartori et al., 2016).

Although the brain alterations leading to acquired paedophilia are evident, they are also spatially heterogeneous, making it unclear which neural network is involved in the onset of the pathological behaviour. The brain regions reported to be altered in offenders with acquired paedophilia include the right orbitofrontal cortex (Burns & Swerdlow, 2003; Fumagalli et al., 2015), the right amygdala (Devinsky et al., 2010), the right globus pallidus (Mendez & Shapira, 2011) the hypothalamus (Frohman et al., 2002; Miller et al., 1986; Sartori et al., 2016), the hippocampus bilaterally (Mendez, 2010; Mendez et al., 2000; Mendez & Shapira, 2011) and the basal ganglia bilaterally (Mendez & Shapira, 2011).

Based on the assumption that each brain region is part of a complex network of regions (Avena-Koenigsberger et al., 2017), a recent study (Scarpazza et al., 2021) identified the regions that are functionally connected to each brain lesion causing paedophilia with the aim to clarify whether the functional impairment of a specific set of brain regions underlies the insurgence of acquired paedophilic behaviour in all the published cases of acquired paedophilia. Of relevance is the finding that the neurological data reveals that the neural bases of acquired paedophilia localize to a common resting state high spatial heterogeneity. Indeed, all the lesions temporally network, despite their associated with acquired paedophilic behaviour are functionally connected with a network involving the orbitofrontal areas, the posterior midline structures, the right inferior temporal gyrus and the left fusiform gyrus. These brain regions are crucial for social cognition (posterior midline structures and right ITG), and in particular for theory of mind (posterior midline structures), emotion recognition (right OFC) and impulse control (right OFC). It is noteworthy that these results match well with the aberrant behaviour pattern in acquired paedophilia, as we describe in the next paragraph related to *modus operandi*. In short, the observation of altered activity in a key region for impulse inhibition fits perfectly with previous evidence from single case descriptions of patients with acquired paedophilia, in whom dis-inhibition was invariably present (Devinsky et al., 2010; Gilbert & Focquaert, 2015; Mendez & Shapira, 2011; Miller et al., 1986; Sartori et al., 2016; Scarpazza et al., 2018c). Moreover, dis-inhibition was recently reported to be a red flag suggesting an acquired origin of paedophilic behaviour (Camperio Ciani et al., 2019), and it explains the dis-organized *modus operandi* of these sexual offenders. Similarly, the observation of altered activity in key regions for social cognition, in particular for theory of mind and emotion recognition, fits well with these patients' inability to understand what is morally wrong with their behaviour (Frohman et al., 2002; Fumagalli et al., 2015; Regestein & Reich, 1978; Sartori et al., 2016; Scarpazza et al., 2018c), even if it this finding is not fully concordant within the literature (Camperio Ciani et al., 2019).

In closing, the aforementioned results on neural basis, revealing subtle and inconsistent brain abnormalities in idiopathic paedophilia versus clear, spatially heterogeneous brain abnormalities that are functionally localized to a specific network of regions in acquired paedophilia, clearly support the emerging idea that the two disorders arguably may reflect distinct neuro-pathologies.

Modus Operandi

The third important difference between idiopathic and acquired paedophilia lies in the *modus operandi* of the offenders. While the *modus operandi* of offenders with idiopathic paedophilia has been described as predatory, the *modus operandi* of acquired paedophilia is usually impulsive.

The *modus operandi* of offenders with idiopathic paedophilia is described as involving predatory, organized and premeditated behaviour. Indeed, idiopathic paedophiles are described to actively search for victims, organize their actions, mask their sexually abusing behaviour, enforce the victim's silence, use psychological and physical violence (Hall & Hall, 2007; Miranda & Corcoran, 2000) and, if caught, might deny their behaviour (Fagan et al., 2002; Hall & Hall, 2007). The earliest studies on the *modus operandi* of idiopathic transgressors provided crucial data to understand the strategies adopted by sexual offenders against children to commit their crimes. For instance, offenders have been found to gradually desensitize their victim to physical contact before moving to sexual touching (Berliner & Conte, 1990; Christensen & Blake, 1990). Moreover, offenders also use some type of coercion and threats (Berliner & Conte, 1990; Budin & Johnson, 1989; Leclerc et al., 2005, 2009). Using a sample of 226 adult offenders, Leclerc et al. (2006) studied the impact of several factors, such as the age of the victim (0–13 years old), on the likelihood of adopting a manipulative, a coercive or a non-persuasive strategy to involve the victim in sexual activity. They found that adult offenders who sexually abuse older children were more likely to use a manipulative, rather than a non-persuasive strategy. Thus, idiopathic paedophilic offences are typically planned in detail.

It is worth noting that some studies identified differences in impulse inhibition between offenders with idiopathic paedophilia and controls, indicating that offenders with idiopathic paedophilia are characterized by weaker impulse control abilities compared to controls (Joyal et al., 2014). However, this result is obtained when comparing a group of offenders with paedophilia with a group of controls, thus making it unclear whether the same results can be translated to each single offender. In any case, the effect size of this effect is so small that the possibility of it being clinically relevant is very unlikely. Finally, some authors also pointed out that so far it is not possible to clarify whether this effect is due to the presence of paedophilia or might be better explained by the co-morbid personality disorders (Mohnke et al., 2014).

A further important detail is that offenders with idiopathic paedophilia perceive their sexual attraction towards children as ego-syntonic (MacMartin & Wood, 2005), meaning that the behaviours, values and feelings are experienced as in harmony with or acceptable to the needs of the individual, or are seen as consistent with one's ideal self-image. For these reasons, offenders with idiopathic paedophilia usually do not portray a sense of guilt.

The *modus operandi* of offenders with acquired paedophilia has been extensively investigated in a recent study (Camperio Ciani et al.,

2019) that, by using an innovative combination of statistical methods, identified the following behavioural red flags with regard to acquired paedophilic behaviour: (1) no evidence of masking the offensive acts; (2) absence of premeditation; (3) spontaneous confession; (4) presence of a sense of guilt. Importantly, two of these behavioural red flags reflect the impulse dis-control that characterize acquired paedophilia (Burns & Swerdlow, 2003; Devinsky et al., 2010): absence of premeditation and no intention to disguise the criminal behaviour, two red flags that are also highly correlated. Indeed, if a behaviour is driven by such *hic and nunc* sexual impulses, it should appear dis-organized. For instance, these offenders assaulted their victim in open spaces, and on occasion even in front of possible witnesses. Furthermore, the selection of their victims is not specifically thought through as a result of the absence of premeditation. For instance, one patient described in the literature abused his own daughter (Rainero et al., 2011), another patient abused his own stepdaughter (Gilbert & Vranic, 2015), another was a paediatrician whom abused his patients in front of the parents (Sartori et al., 2016), and another masturbated in front of a school that was just outside his home (Scarpazza et al., 2018a). These latter behaviours can also be observed in dementia patients or in persons with severe intellectual disability. In other words, they victimized children even if the likelihood to be discovered was very high. Their acts probably reflect the impulse dis-control that characterizes patients with acquired paedophilia (Mohnke et al., 2014) and it could potentially be considered in clinical–anatomical correlation with the structural neural impairments identified in the previous paragraph.

The other two behavioural and/or emotional red flags, spontaneous confession and a sense of guilt, are slightly more difficult to interpret. Both of them might be explained by spared moral judgement that would make the paedophilic behaviour ego-dystonic (Burns & Swerdlow, 2003; Devinsky et al., 2010; Frohman et al., 2002; Solla et al., 2006). However, at least in some cases, the moral judgement of the perpetrator is impaired as well (Lesniak et al., 1972; Sartori et al., 2016; Scarpazza et al., 2018c) and the defendants are not able to understand what is morally wrong with their behaviour. In these cases, they tend to easily confess their crimes as they cannot see anything wrong with them, but a sense of guilt is absent. In one peculiar case (Sartori et al., 2016; Scarpazza et al., 2018a), the defendant was completely incapable to understand the moral disvalue of his acts upon arrest, but a strong sense of guilt emerged after the surgical resection of the tumour.

It is important to underline that the presence of these red flags cannot lead to a clinical diagnosis of acquired paedophilia. Rather, their presence should prompt a rapid neuro-scientific evaluation including at least a brain imaging scan and a comprehensive neurological examination.

Possible Treatments

The fourth important difference between idiopathic and acquired paedophilia lies in the possible treatment options. While idiopathic paedophilia is the primary condition that needs to be treated, acquired paedophilia can theoretically be treated by treating the underlying neurological condition.

There seems to be no evidence to suggest that idiopathic paedophilia can be changed, as is the case with psychiatric disorders such as narcissistic and psychopathic personality disorder, and no treatment is effective unless individuals who are at risk of acting upon their paedophilic urges are willing to engage in treatment (Hall & Hall, 2007; Stone et al., 2000). To minimize the risk of transgressions, psychotherapeutic interventions are designed to increase voluntary control over sexual arousal, reduce sex drive and/or teach self-management skills to individuals who are motivated to avoid acting upon their sexual interests (Seto, 2009). However, despite the fact that psychotherapy is an important aspect of treatment, debate exists concerning its overall effectiveness with regard to the long-term prevention of new offences (Hall & Hall, 2007; Hanson & Morton-Bourgon, 2005; Langton et al., 2006).

Renaud and colleagues (2011) have argued that a real-time functional magnetic imaging (rt-fMRI) brain–computer interface (BCI) neurofeedback system can help to target affected brain areas in individuals with idiopathic paedophilia. Such a system not only may help individuals cultivate behavioural and arousal control associated with their paedophilic urges and ruminations, but may also help provide a form of "covert mental rehearsal" in which unwanted ruminations or impulses are paired with negative stimuli, trained and altered (Johnston et al., 2010; Renaud et al., 2011). Neurofeedback has been suggested as a potentially successful form of treatment, as functional abnormalities associated with idiopathic paedophilia might be related to difficulties in behavioural regulation often observed in diagnosed individuals (Mohnke et al., 2014). However, it is important to highlight that difficulties in behavioural regulation are likely linked to comorbid cognitive–emotional impairments in these offenders rather than the idiopathic paedophilia itself. Another recent pilot programme called the Berlin Dissexuality Therapy (BEDIT) also suggests that forms of cognitive behavioural therapy (CBT) may help individuals with idiopathic paedophilia gain better behavioural and arousal control of their sexual impulses and ruminations (Beier, 2016; Beier et al., 2015). Indeed, CBT can help an individual gain control of prefrontal brain areas over subcortical and other limbic structures known to be associated with paedophilic urges, such as the amygdala (Karlsson, 2011).

To enable effective rehabilitation, psychotherapy is often coupled with androgen deprivation therapy (ADT), by which the individual's testosterone level is lowered to a pre-pubescent level, thereby eliminating or severely reducing sexual urges (Thibaut et al., 2010), or with the administration of selective serotonin reuptake inhibitors as a non-hormonal treatment that has been suggested for paraphilias in general and for paedophilia specifically (Hall & Hall, 2007; Stone et al., 2000). The most promising potential treatment for hindering and controlling the urges and ruminations associated with idiopathic paedophilia may be the use of selective serotonin reuptake inhibitors (SSRIs) as a pharmacological therapy (Gilbert & Outram, 2009). SSRIs, which are commonly used for a variety of psychiatric illnesses, block serotonin reuptake by neurons, leading to amplified serotonin levels in the synaptic gap and the higher likelihood that serotonin will bind to post-synaptic receptors (Frazer, 1997). Genetic research (Berryessa, 2014; Comings, 1994; Tost et al., 2004) and functional imaging research (Schiffer et al., 2017), has suggested the disfunction and involvement of serotonin function and production in the brain in individuals with idiopathic paedophilia. Although no large-scale randomized control trials have studied the efficacy of SSRIs in regulating paedophilic urges, ruminations and behaviours, open label studies and case reports suggest that this type of pharmacological treatment can augment impulse control in those with paedophilia, reduce sexual urges and ruminations and may also dampen sex drive (Bradford, 2001; Kafka, 1994; Kafka & Hennen, 2000; Stein et al., 1992). Nevertheless, after a year of combined psychotherapy and pharmacotherapy, individuals with idiopathic paedophilia still show sexual interest in children, whereas their frequency of urges decreases (Schober et al., 2005), indicating that, while urges can be managed, the core attraction to children does not change (Hall & Hall, 2007; Schober et al., 2005). Also, offenders commonly do not fully comply with psychological and medical treatments (Fagan et al., 2002; Stone et al., 2000), which typically leaves these offenders with a high risk of sexual recidivism (Hanson, 2002; Hanson & Morton-Bourgon, 2005; Seto, 2009; Seto et al., 2004). Hence, in addition to psychotherapy and medical treatments, supervision by family members and/or medical personnel, as well as removing these individuals from toxic or negative environments that may cause reactive offending upon reentry into society need to be considered.

Unlike idiopathic paedophilia, acquired paedophilia can theoretically be addressed by treating the underlying medical (neurological) condition (Sartori et al., 2016). For instance, paedophilia can recede after surgical resection of the tumour causing the paedophilic behaviour (Burns & Swerdlow, 2003; Sartori et al., 2016). So far, no sexual recidivism has been described when individuals with acquired paedophilia received effective treatment for the neurological disorder causing their paedophilic behaviour. To the best of our knowledge, the recurrence of paedophilic urges has only been observed when the neurological disorder itself re-occurred, as explained above when delineating the etiological origin of acquired paedophilia.

Forensic patients	Idiopathic paedophilic disorder	Acquired paedophilia
Type of disorder	Psychiatric	Neurological

(continued)

(continued)

Forensic patients	Idiopathic paedophilic disorder	Acquired paedophilia
Aetiology	Multifactorial	i. Clear aetiology always present ii. Aetiology depends on the underlying neurological disorder
Neural basis	i. No clear structural alterations ii. Subtle structural alterations not consistent across the literature iii. Preliminary findings on functional alterations evident when observing photos of naked bodies of children	i. Brain abnormalities clearly present (lesions of atrophy) ii. Spatial heterogeneity of the brain alterations iii. Brain lesions are functionally connected to a specific brain network
Modus operandi	Premeditated and predatory	Impulsive and disorganized
Possible treatments	Androgen deprivation therapy and/or psychotherapy aimed at reducing paedophilic urges	Medical therapy addressing the underlying neurological condition

Legal Implications for Punishment

In cases where the behaviour of an individual with paedophilia is attributable to acquired brain abnormalities, it is far from obvious that punitive sanctions are the right answer, and treatment rather than punishment is justifiably called for. Indeed, it may be argued that treatment, or at the very least, the option of treatment is the most defensible approach, both from an ethical and a public safety perspective. Retributive punishment tends to increase the risk of recidivism rather than reduce the risk of recidivism. Hence, imposing a retributive sentence on an individual who is successfully treated for his acquired brain abnormality involves the opposite of protecting society (Kelly, 2021). Moreover, surgical resection of the tumour will be a medical necessity in cases where an offender with acquired paedophilia is at immanent risk of dying due to the tumour. Of course, there are less medically urgent cases, in which the individual could similarly be relieved from his deviant thoughts and urges by undergoing brain surgery. And how should we proceed in cases involving perpetrators with Parkinson's disease or Alzheimer's disease who suddenly and unexpectedly commit a sexual offence? How should we approach less straightforward cases involving individuals with idiopathic paedophilia who could potentially be adequately treated by the administration of less invasive neurointerventions such as non-invasive brain stimulation or neurofeedback and/or medications such as androgen deprivation therapy (Focquaert et al., 2020; Gilbert & Focquaert, 2015). Should these offenders be offered such treatments combined with extensive psychotherapy as a community-based rehabilitative alternative to imprisonment? These are difficult questions to answer. Not in the least because the crimes under consideration are deeply immoral and often result in the life-long disruption of the lives of the victims and their loved ones.

Although there are pronounced structural and/or functional neurobiological differences between acquired and idiopathic paedophilia, knowledge on the neural abnormalities of both types of paedophilia may potentially influence legal perspectives on potential criminal sanctions for offenders. Similar to discussions surrounding offenders with psychopathy (Glenn et al., 2009; Levy, 2011; Morse, 2008; Umbach et al., 2015) or behavioural variant frontotemporal dementia (bvFTD) (Berryessa, 2016a), such abnormalities, whether congenital or acquired, may show that traditional sanctions, specifically those aimed at retribution, are ineffective and unjustified punishments for offenders with paedophilia. Instead, alternative sanctions associated with treatment and behavioural control may be more appropriate in addressing offending behaviour by both populations (Berryessa, 2014, 2021), both in the interest of society and the resocialization and rehabilitation of the offender (Focquaert et al., 2020).

Retribution, also described as "just deserts" punishment, is often considered one of, if not the, main objective of Anglo-American criminal law. Retributive sentences are given in accordance with what is believed to be an appropriate or deserved punishment for a criminal offender based on his moral responsibility for his actions, and may be combined with incapacitating sentences as well to ensure public safety (Smith,

2005). Culturally, sex offenders with paedophilia are viewed as "the worst of the worst", and are often subjected to severe, retributive punishments, lengthy periods of incarceration for their crimes, and "shaming" post-conviction policies upon their release (Nhan et al., 2012; Quinn et al., 2004). Such legal practices are and have been directly influenced by the immense societal stigmas that paedophilic behaviour carries (Berryessa & Lively, 2019; Bumby & Maddox, 1999). Although such social and legal perspectives towards paedophilia are persistent and pervasive, the structural and/or functional neurological impairments associated with both acquired and idiopathic paedophilia may help provide evidence to counter stigmas on the origins and ruminations associated with paedophilia and that retributive punishments are unlikely to "solve the problem" related to future paedophilic behaviour. Instead, alternative strategies may be needed to hinder future offending by individuals with both idiopathic and acquired paedophilia.

Idiopathic Paedophilia

Judges may view research on its brain abnormalities as evidence that offenders with paedophilia are pathologically and congenitally disordered, and although legally culpable, may be less morally responsible for their actions due their "broken brains" (Berryessa, 2021; Monterosso et al., 2005). For example, a judge may consider brain abnormalities associated with idiopathic paedophilia as potential contributors to an offender's sexual crimes and his difficulties in emotionally and morally evaluating his actions. This may not only assuage retributive sentiments of the court, which are based on the perceived "moral desert" of an offender for his actions, but may open up support for more treatment oriented sentences aimed at treating the frontal and subcortical deficits associated with paedophilia that may help to impede ruminations and actions, associated with paedophilic offending. Although changes in punishment perspectives for offenders with paedophilia would rely on judges being open and receptive to such neuroscience evidence, there is some evidence to show that judges are open to the consideration of biological evidence on behaviour as potential mitigators to sentencing practices and may even foster support for more treatment oriented sanctions (Berryessa, 2016b; Moriarty, 2008). Indeed, lay experimental samples have also viewed biological evidence on paedophilia as mitigating to prison sentences (Berryessa, 2018).

Instead of retributive punishments, rehabilitative treatments that target the functional neurological abnormalities associated with idiopathic may be more effective methods to prevent future offending, particularly as a counter to existing retributive policies targeted towards sexual offending by individuals with paedophilia that are largely ineffective in preventing sexual recidivism (Tewksbury & Lees, 2007). Indeed, the best potential treatments to help prevent future offending should likely be aimed at regulating and affecting the neural deficits associated with idiopathic paedophilia and helping individuals gain better behavioural and emotional control over their sexual arousal and attain self-management skills in order to regulate their sexual urges (Berryessa, 2021). Therefore, more research is urgently needed to identify the neural deficits in sexual offenders with idiopathic paedophilia and the extent to which these are related to or interact with comorbid disorders.

As mentioned above, recent literature has discussed potential treatment options for idiopathic paedophilia as potential future rehabilitative sentencing alternatives (see Berryessa, 2021). Although future research is necessary on these and other treatments, they do represent promising rehabilitative treatments option that might be effective in preventing future offending by individuals with idiopathic paedophilia by regulating and altering, rather than retributively punishing, the actions associated with this type of offending. Research on preventative tools or strategies to truly affect sexual recidivism may help to save both future child victimization as well as valuable monetary and social resources that we currently devote to largely ineffective sexual offender policies and punishments (Berryessa, 2021; Tewksbury & Lees, 2007). Moreover, in addition to applicable non-retributive legal sanctions and rehabilitative trajectories, restorative justice practices need to be offered and made available to the victims and their loved ones (Focquaert, 2020; Johnstone, 2021).

Acquired Paedophilia

The neurological and behavioural changes associated with the development of acquired paedophilia, paired with the substantial deficits to empathy, normal emotional processing, moral decision-making and in following legal and moral norms exhibited by diagnosed individuals, also suggest that retributive sentences appear to be ineffective methods of punishment for those with acquired paedophilia. Offenders with acquired paedophilia, like those with behavioural variant frontotemporal dementia (bvFTD), can rationally recognize that their actions may be considered morally or legally wrong, but do not have the neurological capacity for true moral understanding of their actions (Mendez & Shapira, 2009). Even if a person exhibited normal mores and decision-making before the neurological changes associated with acquired paedophilia, an individual's ability to practice or understand socio-moral behaviour becomes compromised as the disease develops (Burns & Swerdlow, 2003). As such, for similar reasons noted above, one must consider whether an individual with acquired paedophilia may have the capacity for true moral responsibility, and correspondingly, whether retributive punishments based upon it are warranted (Gilbert & Focquaert, 2015).

Further, traditional punishments that focus on deterrent strategies in order to prevent future offending may also not be appropriate or effective sentences for offenders with acquired paedophilia. Deterrent sentences rely on offenders recognizing the risk or costs of potential punishment to stop offending behaviour (Smith, 2005). However, similar to some other neurological disorders such as bvFTD and psychopathy, individuals with acquired paedophilia often have structural and functional impairments in brain areas involved in the neural circuitry underlying punishment association, which may lead to less sensitivity to punishment (Mitchell et al., 2006; Rosen et al., 2002; Sturm et al., 2017). Evidence on deficits to neural circuitry underlying punishment association suggests that individuals with such neural abnormalities may be insensitive to deterrent sentences and fears of future punishment (Aharoni et al., 2007). Thus, individuals with acquired paedophilia may be undeterred by existing sentencing strategies aimed at retributive or deterrent objectives.

Instead, like those with idiopathic paedophilia, offenders with acquired paedophilia may be more likely to benefit from alternatives to traditional punishments. As the underlying neural abnormalities differ between both types of paedophilia, potential treatment options may differ from those suggested for idiopathic paedophilia. Even so, several brain regions involved in acquired paedophilia are also related to diminished behavioural control (Mohnke et al., 2014). Thus, medications and similar treatment programmes to those described above for idiopathic paedophilia, including SSRIs, could also be effective (Tsai & Boxer, 2014).

Yet, there are some behavioural differences between acquired and idiopathic paedophilia that may suggest some unique alterative sentences for those with acquired paedophilia. Individuals with acquired paedophilia do not exhibit goal-oriented or premeditated behaviour in offending, and instead, are more often reacting directly in response to stimuli in their immediate environment (Gilbert & Focquaert, 2015; Sartori et al., 2016). Thus, it is possible that removing individuals with acquired paedophilia from an environment that may provide stimuli for offending behaviour likely will decrease the likelihood of future paedophilic behaviour. This has been true for individuals with other neurological disorders with similar neural deficits. For example, (Warren et al., 2013) have argued that supervision of individuals with bvFTD by family members or medical personnel, as well as removal from toxic or negative environments that may cause reactive offending, is effective in hindering behavioural and moral impairments associated with bvFTD. Regulating the environments of those with acquired paedophilia likely can help contain offending without more serious forms of punishment. Prison environments are typically criminogenic rather than conducive to the protection of society as incarceration delays and often problematizes desistance substantially (Canton, 2017; Kelly, 2021). At the same time, for certain offenders with paedophilia, especially in case of comorbidity with antisocial and psychopathic traits, the risk of committing additional sexual crimes in combination with treatment-resistance/refusal will not warrant their reentry into society.

Ethical Implications Related to Treatment in Offenders with Idiopathic Paedophilia

If biomedical psychiatric treatments aimed at regulating and affecting the neural deficits associated with idiopathic paedophilia could adequately reduce the risk of sexual transgressions in offenders with idiopathic paedophilia, should the criminal justice system be permitted to use such biomedical interventions and, if so, under which conditions? For example, the Netherlands allows a judge to impose a prison sentence followed by postprison mandated forensic psychiatric treatment, either residential or nonresidential, of mentally ill offenders provided that such treatment has been deemed necessary at the time of sentencing. For nonviolent crimes, the maximum duration is four years after the offender has completed his prison sentence. For violent crimes, the duration may be extended (indefinitely) if certain conditions are met. Provided the length of the entire legal mandate does not amount to cases of dual jeopardy, such postprison treatment, possibly involving neurointerventions such as psychopharmacological treatment, might provide the most defensible criminal justice approach for individuals with idiopathic paedophilia in comparison to a lengthy prison sentence without any type of treatment, rehabilitation or resocialization. Importantly, two things are essential for this to work: (a) the offender's willingness to cooperate in the treatment; and (b) effective treatment options.

Focquaert et al. (2020) have argued that non-retributive criminal justice approaches in which neurointerventions are offered to offenders as a condition of probation, parole, or sentence reduction *can* be ethical provided that the following minimal ethical conditions are met: (a) the status quo is in no way cruel, inhuman, degrading or in some other way *wrong* (the status quo being the alternative to the offer; e.g. the nature of detention if one declines the offer); (b) the neurointervention itself is in no way cruel, inhuman, degrading, or in some other way *wrong*; (c) the neurointervention respects the well-being of the offender; (d) the neurointervention targets one or more risk factors for recidivism; and (e) the neurointervention is voluntary: the offender is formally required to give his or her free and informed consent upon acceptance, and, if appropriate, a court-appointed guardian his or her authorization. The latter condition requires that the offender gives his or her free and informed consent upon acceptance, or, where the person concerned is not deemed to be legally competent, authorization is provided by that person's legal representative, the offender does not object, and the offender takes part as far as possible in the authorization procedure.

If offenders with idiopathic paedophilia are willing to cooperate in treatment and effective treatment options exist, the failure of our current punishment practices in reducing recidivism necessitate an academic, societal and political debate where the implementation of such alternative non-retributive approaches to crime is considered a viable component of the much needed reform of our current punishment practices (Focquaert et al., 2021).

Conclusion

Although there are striking differences between acquired and idiopathic paedophilia in terms of the neurological correlates, aetiology and treatment options of these disorders, research on the structural and/or functional neurological impairments associated with both acquired and idiopathic paedophilia may help counter stigmas on the origins and ruminations associated with pedophilia.

As we outlined in detail in this chapter, the aetiology of idiopathic paedophilic disorder is still unknown, but is considered to be multifactorial, encompassing biological, psychological, and social factors. In contrast, acquired paedophilic behaviour refers to the insurgence of sexual urges towards children later in life as a consequence of an acquired neurological condition with a clear neurologic aetiology. Moreover, while idiopathic paedophilia is the primary condition that needs to be treated, acquired paedophilia can, at least in theory, be treated by addressing the underlying neurological condition. Treating idiopathic paedophilia is a complex matter, that may involve psychotherapy, pharmacological therapy and the need for a supervision system. Knowledge on the functional and/or structural deficits in idiopathic paedophilia may enable the development of more effective treatments.

Based on our current scientific knowledge regarding both disorders, retributive punishments are unlikely to "solve the problem" related to future paedophilic behaviour. Instead, alternative strategies may be needed to hinder future offending by individuals with both idiopathic and acquired paedophilia. If an alternative non-retributive route is considered, it is imperative that in addition to any applicable non-retributive legal sanctions and/or rehabilitative trajectories, restorative justice practices and psychological counselling are immediately made available to the victims and their loved ones.

References

Aharoni, E., Weintraub, L. L., & Fridlund, A. J. (2007). No skin off my back: Retribution deficits in psychopathic motives for punishment. *Behavioral Sciences & the Law, 25*(6), 869–889. https://doi.org/10.1002/bsl.790.

Avena-Koenigsberger, A., Misic, B., & Sporns, O. (2017). Communication dynamics in complex brain networks. *Nature Reviews Neuroscience, 19*(1), 17–33. https://doi.org/10.1038/nrn.2017.149.

Beech, A. R., Miner, M. H., & Thornton, D. (2016). Paraphilias in the DSM-5. *Annual Review of Clinical Psychology, 12,* 383–406. https://doi.org/10.1146/annurev-clinpsy-021815-093330.

Beier, K. M. (2016). Proactive strategies to prevent child sexual abuse and the use of child abuse images: Experiences from the German Dunkelfeld project. *Women and children as victims and offenders: Background, prevention, reintegration* (pp. 499–524). Cham: Springer.

Beier, K. M., Grundmann, D., Kuhle, L. F., Scherner, G., Konrad, A., & Amelung, T. (2015). The German Dunkelfeld project: A pilot study to prevent child sexual abuse and the use of child abusive images. *The Journal of Sexual Medicine, 12*(2), 529–542. https://doi.org/10.1111/jsm.12785.

Berliner, L., & Conte, J. R. (1990). The process of victimization: The victims' perspective. *Child Abuse and Neglect, 14*(1), 29–40. https://doi.org/10.1016/0145-2134(90)90078-8.

Berryessa, C. M. (2014). Potential implications of research on genetic or heritable contributions to pedophilia for the objectives of criminal law. *Recent*

Advances in DNA & Gene Sequences, 8(2), 65–77. https://doi.org/10.2174/2352092209666141211233857.
Berryessa, C. M. (2016). Behavioral and neural impairments of frontotemporal dementia: Potential implications for criminal responsibility and sentencing. *International Journal of Law and Psychiatry, 46,* 1–6. https://doi.org/10.1016/j.ijlp.2016.02.020.
Berryessa, C. M. (2016). Judges' views on evidence of genetic contributions to mental disorders in court. *J Forens Psychiatry Psychol, 27*(4), 586–600. https://doi.org/10.1080/14789949.2016.1173718.
Berryessa, C. M. (2018). The effects of psychiatric and "biological" labels on lay sentencing and punishment decisions. *Journal of Experimental Criminology, 14*(2), 241–256.
Berryessa, C. M. (2021). Brain abnormalities associated with pedophilic disorder: Implications for retribution and rehabilitation. In F. Focquaert, E. Shaw, & B. N. Waller (Eds.), *The Routledge handbook of the philosophy and science of punishment* (pp. 231–254). London, UK: Routledge.
Berryessa, C. M., & Lively, C. (2019). When a sex offender wins the lottery: Social and legal punitiveness toward sex offenders in an instance of perceived injustice. *Psychology, Public Policy, and Law, 25*(3), 181–195.
Bradford, J. M. (2001). The neurobiology, neuropharmacology, and pharmacological treatment of the paraphilias and compulsive sexual behaviour. *Canadian Journal of Psychiatry, 46*(1), 26–34. https://doi.org/10.1177/070674370104600104.
Budin, L. E., & Johnson, C. F. (1989). Sex abuse prevention programs: Offenders' attitudes about their efficacy. *Child Abuse and Neglect, 13,* 77–88.
Bumby, K. M., & Maddox, M. C. (1999). Judges' knowledge about sexual offenders, difficulties presiding over sexual offense cases, and opinions on sentencing, treatment, and legislation. *Sex Abuse, 11*(4), 305–315. https://doi.org/10.1177/107906329901100406.
Burns, J. M., & Swerdlow, R. H. (2003). Right orbitofrontal tumor with pedophilia symptom and constructional apraxia sign. *Archives of Neurology, 60*(3), 437–440.
Butler, C., & Zeman, A. Z. (2005). Neurological syndromes which can be mistaken for psychiatric conditions. *Journal of Neurology, Neurosurgery and Psychiatry, 76*(Suppl. 1), i31–i38. https://doi.org/10.1136/jnnp.2004.060459.
Camperio Ciani, A. S., Scarpazza, C., Covelli, V., & Battaglia, U. (2019). Profiling acquired pedophilic behavior: Retrospective analysis of 66 Italian

forensic cases of pedophilia. *International Journal of Law and Psychiatry, 67,* 101508. https://doi.org/10.1016/j.ijlp.2019.101508.
Canton, R. (2017). *Why punish? An introduction to the philosophy of punishment.* New York, NY: Palgrave.
Cantor, J. M., Kabani, N., Christensen, B. K., Zipursky, R. B., Barbaree, H. E., Dickey, R., Klassen, P. E., Mikulis, D. J., Kuban, M. E., Blak, T., Richards, B. A., Hanratty, M. K., & Blanchard, R. (2008). Cerebral white matter deficiencies in pedophilic men. *Journal of Psychiatric Research, 42*(3), 167–183. https://doi.org/10.1016/j.jpsychires.2007.10.013.
Christensen, J. R., & Blake, R. H. (1990). The grooming process in father-daughter incest. In A. L. Horton, B. L. Johnson, L. M. Rowndy, & D. Williams (Eds.), *The incest perpetrator: A family member no one wants to treat* (pp. 88–98). Sage.
Cohen, L. J., & Galynker, I. I. (2002). Clinical features of pedophilia and implications for treatment. *Journal of Psychiatric Practice, 8*(5), 276–289.
Comings, D. E. (1994). Genetic factors in substance abuse based on studies of Tourette syndrome and ADHD probands and relatives. I. Drug abuse. *Drug and Alcohol Dependence, 35*(1), 1–16. https://doi.org/10.1016/0376-8716(94)90104-x.
Devinsky, J., Sacks, O., & Devinsky, O. (2010). Kluver-Bucy syndrome, hypersexuality, and the law. *Neurocase, 16*(2), 140–145. https://doi.org/10.1080/13554790903329182.
Doshi, S. M., Zanzrukiya, K., & Kumar, L. (2018). Paraphilic infantilism, diaperism and pedophilia: A review. *Journal of Forensic and Legal Medicine, 56,* 12–15. https://doi.org/10.1016/j.jflm.2018.02.026.
Eher, R., Rettenberger, M., & Turner, D. (2019). The prevalence of mental disorders in incarcerated contact sexual offenders. *Acta Psychiatrica Scandinavica, 139*(6), 572–581. https://doi.org/10.1111/acps.13024.
Fagan, P. J., Wise, T. N., Schmidt, C. W., Jr., & Berlin, F. S. (2002). Pedophilia. *JAMA, 288*(19), 2458–2465. https://doi.org/10.1001/jama.288.19.2458.
Focquaert, F. (2020). Free will skepticism and punishment: A preliminary ethical analysis. In E. Shaw, D. Pereboom, & G. Caruso (Eds.), *Free will skepticism in law and society: Challenging retributive justice* (pp. 207–236). Cambridge University Press.
Focquaert, F., Shaw, E., Waller, B.N. (Eds.). (2021). *The routledge handbook of the philosophy and science of punishment*. New York and London: Routledge Taylor & Francis Group.
Focquaert, F., Van Assche, K., & Sterckx, S. (2020). Offering neurointerventions to offenders with cognitive-emotional impairments: Ethical and

criminal justice aspects. In N. A. Vincent, T. Nadelhoffer, & A. McCay (Eds.), *Neurointerventions and the law. Regulating human mental capacity* (pp. 128–149). Oxford University Press.

Frazer, A. (1997). Pharmacology of antidepressants. *Journal of Clinical Psychopharmacology, 17*(Suppl. 1), 2S–18S. https://doi.org/10.1097/00004714-199704001-00002.

Freund, K., & Kuban, M. (1994). The basis of the abused abuser theory of pedophilia: A further elaboration on an earlier study. *Archives of Sexual Behavior, 23*(5), 553–563. https://doi.org/10.1007/bf01541497.

Freund, K., Watson, R., & Dickey, R. (1990). Does sexual abuse in childhood cause pedophilia: An exploratory study. *Archives of Sexual Behavior, 19*(6), 557–568. https://doi.org/10.1007/bf01542465.

Frohman, E. M., Frohman, T. C., & Moreault, A. M. (2002). Acquired sexual paraphilia in patients with multiple sclerosis. *Archives of Neurology, 59*(6), 1006–1010.

Fumagalli, M., Pravettoni, G., & Priori, A. (2015). Pedophilia 30 years after a traumatic brain injury. *Neurological Sciences, 36*(3), 481–482. https://doi.org/10.1007/s10072-014-1915-1.

Geer, J. H., Estupinan, L. A., & Manguno-Mire, G. M. (2000). Empathy, social skills, and other relevant cognitive processes in rapists and child molesters. *Aggression and Violent Behavior, 5,* 99–126.

Gilbert, F., & Focquaert, F. (2015). Rethinking responsibility in offenders with acquired paedophilia: Punishment or treatment? *International Journal of Law and Psychiatry, 38,* 51–60. https://doi.org/10.1016/j.ijlp.2015.01.007.

Gilbert, F., & Outram, S. (2009, September). Chemical interventions and ethical side effects. *Canadian Chemical News (L'Actualité Chimique Canadienne),* 20–21.

Gilbert, F., & Vranic, A. (2015). Paedophilia, invasive brain surgery, and punishment. *Journal of Bioethical Inquiry, 12*(3), 521–526. https://doi.org/10.1007/s11673-015-9647-3.

Glenn, A. L., Raine, A., & Schug, R. A. (2009). The neural correlates of moral decision-making in psychopathy. *Molecular Psychiatry, 14*(1), 5–6. https://doi.org/10.1038/mp.2008.104.

Goodkind, M., Eickhoff, S. B., Oathes, D. J., Jiang, Y., Chang, A., Jones-Hagata, L. B., Ortega, B. N., Zaiko, Y. V., Roach, E. L., Korgaonkar, M. S., Grieve, S. M., Galatzer-Levy, I., Fox, P. T., & Etkin, A. (2015). Identification of a common neurobiological substrate for mental illness. *JAMA Psychiatry, 72*(4), 305–315. https://doi.org/10.1001/jamapsychiatry.2014.2206.

Green, R. (2002). Is pedophilia a mental disorder? *Archives of Sexual Behavior, 31*(6), 467–471; discussion 479–510. https://doi.org/10.1023/a:1020699013309.

Greenberg, D. M., Bradford, J. M., & Curry, S. (1993). A comparison of sexual victimization in the childhoods of pedophiles and hebephiles. *Journal of Forensic Sciences, 38*(2), 432–436.

Hall, R. C., & Hall, R. C. (2007). A profile of pedophilia: Definition, characteristics of offenders, recidivism, treatment outcomes, and forensic issues. *Mayo Clinic Proceedings, 82*(4), 457–471. https://doi.org/10.4065/82.4.457.

Hanson, R. K. (2002). Recidivism and age—Follow-up data from 4,673 sexual offenders. *Journal of Interpersonal Violence, 17*(10), 1046–1062. https://doi.org/10.1177/088626002236659.

Hanson, R. K., & Morton-Bourgon, K. E. (2005). The characteristics of persistent sexual offenders: A meta-analysis of recidivism studies. *Journal of Consulting and Clinical Psychology, 73*(6), 1154–1163. https://doi.org/10.1037/0022-006X.73.6.1154.

Hanson, R. K., Morton, K. E., & Harris, A. J. (2003). Sexual offender recidivism risk: What we know and what we need to know. *Annals of the New York Academy of Sciences, 989*, 154–166; discussion 236–146.

Hanson, R. K., Steffy, R. A., & Gauthier, R. (1993). Long-term recidivism of child molesters. *Journal of Consulting and Clinical Psychology, 61*(4), 646–652. https://doi.org/10.1037/0022-006x.61.4.646.

Jespersen, A. F., Lalumiere, M. L., & Seto, M. C. (2009). Sexual abuse history among adult sex offenders and non-sex offenders: A meta-analysis. *Child Abuse and Neglect, 33*(3), 179–192. https://doi.org/10.1016/j.chiabu.2008.07.004.

Johnston, S. J., Boehm, S. G., Healy, D., Goebel, R., & Linden, D. E. (2010). Neurofeedback: A promising tool for the self-regulation of emotion networks. *Neuroimage, 49*(1), 1066–1072. https://doi.org/10.1016/j.neuroimage.2009.07.056.

Johnstone, G. (2021). The restorative justice movement: Questioning the rationale of contemporary criminal justice. In F. Focquaert, E. Shaw, & B. Waller (Eds.), *The Routledge handbook of the philosophy and science of punishment* (pp. 75–86). London: Routledge.

Joyal, C. C., Beaulieu-Plante, J., & de Chantérac, A. (2014). The neuropsychology of sex offenders: A meta-analysis. *Sexual Abuse: A Journal of Research and Treatment, 26*(2), 149–177.

Kafka, M. P. (1994). Sertraline pharmacotherapy for paraphilias and paraphilia-related disorders: An open trial. *Annals of Clinical Psychiatry, 6*(3), 189–195. https://doi.org/10.3109/10401239409149003.

Kafka, M. P., & Hennen, J. (2000). Psychostimulant augmentation during treatment with selective serotonin reuptake inhibitors in men with paraphilias and paraphilia-related disorders: A case series. *Journal of Clinical Psychiatry, 61*(9), 664–670. https://doi.org/10.4088/jcp.v61n0912.

Karlsson, H. (2011). How psychotherapy changes the brain: Understanding the mechanisms. *Psychiatric Times, 28*(8), 21–21.

Kelly, W. R. (2021). Punishment and its alternatives. In F. Focquaert, E. Shaw, & B. Waller (Eds.), *The Routledge handbook of the philosophy and science of punishment* (pp. 333–343). London: Routledge.

Keshavan, M. S., & Kaneko, Y. (2013). Secondary psychoses: An update. *World Psychiatry, 12*(1), 4–15. https://doi.org/10.1002/wps.20001.

Kruger, T. H., & Schiffer, B. (2011). Neurocognitive and personality factors in homo- and heterosexual pedophiles and controls. *The Journal of Sexual Medicine, 8*(6), 1650–1659. https://doi.org/10.1111/j.1743-6109.2009.01564.x.

Kruger, T. H. C., Sinke, C., Kneer, J., Tenbergen, G., Khan, A. Q., Burkert, A., Müller-Engling, L., Engler, H., Gerwinn, H., von Wurmb-Schwark, N., Pohl, A., Weiß, S., Amelung, T., Mohnke, S., Massau, C., Kärgel, C., Walter, M., Schiltz, K., Beier, K. M., … Frieling, H. (2019). Child sexual offenders show prenatal and epigenetic alterations of the androgen system. *Translational Psychiatry, 9*(1), 28. https://doi.org/10.1038/s41398-018-0326-0.

Langton, C. M., Barbaree, H. E., Harkins, L., & Peacock, E. J. (2006). Sex offenders' response to treatment and its association with recidivism as a function of psychopathy. *Sex Abuse, 18*(1), 99–120. https://doi.org/10.1177/107906320601800107.

Leclerc, B., Carpentier, J., & Proulx, J. (2006). Strategies adopted by sexual offenders to involve children in sexual activity. In R. Wortley & S. Smallbone (Eds.), *Situational prevention of child sexual abuse*. Crime Prevention Studies, Vol. 19 (pp. 251−270). Monsey, NY: Criminal Justice Press.

Leclerc, B., Proulx, J., & Beauregard, E. (2009). Examining the modus operandi of sexual offenders against children and its practical implications. *Aggression and Violent Behavior, 14*(1), 5–12.

Leclerc, B., Proulx, J., & McKibben, A. (2005). Modus operandi of sexual offenders working or doing voluntary work with children and adolescents. *Journal of Sexual Aggression, 2*, 99–120.

Lesniak, R., Szymusik, A., & Chrzanowski, R. (1972). Case report: Multidirectional disorders of sexual drive in a case of brain tumour. *Forensic Science, 1*(3), 333–338.

Levy, K. (2011). Dangerous psychopaths: Criminally responsible but not morally responsible, subject to criminal punishment and to preventive detention. *San Diego Law Review, 48,* 1299–1395.

MacMartin, C., & Wood, L. A. (2005). Sexual motives and sentencing—Judicial discourse in cases of child sexual abuse. *Journal of Language and Social Psychology, 24*(2), 139–159.

Mendez, M., & Shapira, J. S. (2011). Pedophilic behavior from brain disease. *The Journal of Sexual Medicine, 8*(4), 1092–1100. https://doi.org/10.1111/j.1743-6109.2010.02172.x.

Mendez, M. F. (2010). The unique predisposition to criminal violations in frontotemporal dementia. *Journal of the American Academy of Psychiatry, 38*(3), 318–323.

Mendez, M. F., Chow, T., Ringman, J., Twitchell, G., & Hinkin, C. H. (2000). Pedophilia and temporal lobe disturbances. *Journal of Neuropsychiatry and Clinical Neurosciences, 12*(1), 71–76. https://doi.org/10.1176/jnp.12.1.71.

Mendez, M. F., & Shapira, J. S. (2009). Altered emotional morality in frontotemporal dementia. *Cognitive Neuropsychiatry, 14*(3), 165–179. https://doi.org/10.1080/13546800902924122.

Miller, B. L., Cummings, J. L., McIntyre, H., Ebers, G., & Grode, M. (1986). Hypersexuality or altered sexual preference following brain injury. *Journal of Neurology, Neurosurgery and Psychiatry, 49*(8), 867–873. https://doi.org/10.1136/jnnp.49.8.867.

Miranda, A. O., & Corcoran, C. L. (2000). Comparison of perpetration characteristics between male juvenile and adult sexual offenders: Preliminary results. *Sexual Abuse, 12*(3), 179–188. https://doi.org/10.1177/107906320001200302.

Mitchell, D. G., Avny, S. B., & Blair, R. J. (2006). Divergent patterns of aggressive and neurocognitive characteristics in acquired versus developmental psychopathy. *Neurocase, 12*(3), 164–178. https://doi.org/10.1080/13554790600611288.

Mohnke, S., Muller, S., Amelung, T., Kruger, T. H., Ponseti, J., Schiffer, B., Walter, M., Beier, K. M., & Walter, H. (2014). Brain alterations in paedophilia: A critical review. *Progress in Neurobiology, 122,* 1–23. https://doi.org/10.1016/j.pneurobio.2014.07.005.

Monterosso, J., Royzman, E. B., & Schwartz, B. (2005). Explaining away responsibility: Effects of scientific explanation on perceived culpability. *Ethics & Behavior, 15*(2), 139–158.

Moriarty, J. C. (2008). Flickering admissibility: Neuroimaging evidence in the U.S. courts. *Behavioral Sciences & the Law, 26*(1), 29–49. https://doi.org/10.1002/bsl.795.

Morse, S. J. (2008). Psychopathy and criminal responsibility. *Neuroethics, 1*(3), 205–212.

Nhan, J., Polzer, K., & Ferguson, J. (2012). "More dangerous than hitmen": Judicial perceptions of sexual offenders. *International Journal of Criminology and Sociological Theory, 5*(1), 823–836.

Poeppl, T. B., Nitschke, J., Santtila, P., Schecklmann, M., Langguth, B., Greenlee, M. W., Osterheider, M., & Mokros, A. (2013). Association between brain structure and phenotypic characteristics in pedophilia. *Journal of Psychiatric Research, 47*(5), 678–685. https://doi.org/10.1016/j.jpsychires.2013.01.003.

Ponseti, J., Granert, O., Jansen, O., Wolff, S., Beier, K., Neutze, J., Deuschl, G., Mehdorn, H., Siebner, H., & Bosinski, H. (2012). Assessment of pedophilia using hemodynamic brain response to sexual stimuli. *Archives of General Psychiatry, 69*(2), 187–194. https://doi.org/10.1001/archgenpsychiatry.2011.130.

Quinn, J. F., Forsyth, C. J., & Mullen-Quinn, C. (2004). Societal reaction to sex offenders: A review of the origins and results of the myths surrounding their crimes and treatment amenability. *Deviant Behavior, 25*(3), 215–232.

Rainero, I., Rubino, E., Negro, E., Gallone, S., Galimberti, D., Gentile, S., Scarpini, E., & Pinessi, L. (2011). Heterosexual pedophilia in a frontotemporal dementia patient with a mutation in the progranulin gene. *Biological Psychiatry, 70*(9), e43–e44. https://doi.org/10.1016/j.biopsych.2011.06.015.

Raymond, N. C., Coleman, E., Ohlerking, F., Christenson, G. A., & Miner, M. (1999). Psychiatric comorbidity in pedophilic sex offenders. *American Journal of Psychiatry, 156*(5), 786–788. https://doi.org/10.1176/ajp.156.5.786.

Regestein, Q. R., & Reich, P. (1978). Pedophilia occurring after onset of cognitive impairment. *The Journal of Nervous and Mental Disease, 166*(11), 794–798.

Renaud, P., Joyal, C., Stoleru, S., Goyette, M., Weiskopf, N., & Birbaumer, N. (2011). Real-time functional magnetic imaging—Brain–computer interface and virtual reality: Promising tools for the treatment of pedophilia. *Progress in Brain Research, 192,* 263–272.

Rosen, H. J., Gorno-Tempini, M. L., Goldman, W. P., Perry, R. J., Schuff, N., Weiner, M., Feiwell, R., Kramer, J. H., & Miller, B. L. (2002). Patterns of brain atrophy in frontotemporal dementia and semantic dementia. *Neurology, 58*(2), 198–208. https://doi.org/10.1212/wnl.58.2.198.

Sartori, G., Scarpazza, C., Codognotto, S., & Pietrini, P. (2016). An unusual case of acquired pedophilic behavior following compression of orbitofrontal cortex and hypothalamus by a Clivus Chordoma. *Journal of Neurology, 263*(7), 1454–1455. https://doi.org/10.1007/s00415-016-8143-y.

Scarpazza, C., Finos, L., Genon, S., Masiero, L., Bortolato, E., Cavaliere, C., Pezzaioli, J., Monaro, M., Navarin, N., Battaglia, U., Pietrini, P., Ferracuti, S., Sartori, G., & Camperio Ciani, A. S. (2021, January 28). Idiopathic and acquired pedophilia as two distinct disorders: An insight from neuroimaging. *Brain Imaging Behav.* https://doi.org/10.1007/s11682-020-00442-z.

Scarpazza, C., Pellegrini, S., Pietrini, P., & Sartori, G. (2018a). The role of neuroscience in the evaluation of mental insanity: On the controversies in Italy. *Neuroethics, 11*(1), 83–95.

Scarpazza, C., Pellegrini, S., Pietrini, P., & Sartori, G. (2018b). The role of neuroscience in the evaluation of mental insanity: On the controversies in Italy: Comment on "On the stand. Another episode of neuroscience and law discussion from Italy". *Neuroethics, 11*(1), 83–95.

Scarpazza, C., Pennati, A., & Sartori, G. (2018c). Mental insanity assessment of pedophilia: The importance of the trans-disciplinary approach. Reflections on Two Cases. *Front Neurosci, 12,* 335. https://doi.org/10.3389/fnins.2018.00335.

Schiffer, B., Amelung, T., Pohl, A., Kaergel, C., Tenbergen, G., Gerwinn, H., Mohnke, S., Massau, C., Matthias, W., Weiß, S., Marr, V., Beier, K. M., Walter, M., Ponseti, J., Krüger, T. H. C., Schiltz, K., & Walter, H. (2017). Gray matter anomalies in pedophiles with and without a history of child sexual offending. *Translational Psychiatry, 7*(5), e1129. https://doi.org/10.1038/tp.2017.96.

Schiffer, B., Peschel, T., Paul, T., Gizewski, E., Forsting, M., Leygraf, N., Schedlowski, M., & Krueger, T. H. (2007). Structural brain abnormalities in the frontostriatal system and cerebellum in pedophilia. *Journal of Psychiatric Research, 41*(9), 753–762. https://doi.org/10.1016/j.jpsychires.2006.06.003.

Schiltz, K., Witzel, J., Northoff, G., Zierhut, K., Gubka, U., Fellmann, H., Kaufmann, J., Tempelmann, C., Wiebking, C., & Bogerts, B. (2007). Brain pathology in pedophilic offenders: Evidence of volume reduction in the right amygdala and related diencephalic structures. *Archives of General Psychiatry, 64*(6), 737–746. https://doi.org/10.1001/archpsyc.64.6.737.

Schober, J. M., Kuhn, P. J., Kovacs, P. G., Earle, J. H., Byrne, P. M., & Fries, R. A. (2005). Leuprolide acetate suppresses pedophilic urges and arousability. *Archives of Sexual Behavior, 34*(6), 691–705. https://doi.org/10.1007/s10508-005-7929-2.

Seto, M. C. (2009). Pedophilia. *Annual Review of Clinical Psychology, 5,* 391–407. https://doi.org/10.1146/annurev.clinpsy.032408.153618.

Seto, M. C., Harris, G. T., Rice, M. E., & Barbaree, H. E. (2004). The screening scale for pedophilic interests predicts recidivism among adult sex offenders with child victims. *Archives of Sexual Behavior, 33*(5), 455–466. https://doi.org/10.1023/B:ASEB.0000037426.55935.9c.

Sha, Z., Wager, T. D., Mechelli, A., & He, Y. (2019). Common dysfunction of large-scale neurocognitive networks across psychiatric disorders. *Biological Psychiatry, 85*(5), 379–388. https://doi.org/10.1016/j.biopsych.2018.11.011.

Smith, N. (2005). Punishment. In P. Gerstenfeld (Ed.), *Criminal justice encyclopedia* (pp. 894–900). Pasadena, CA: Salem Press.

Solla, P., Floris, G., Tacconi, P., & Cannas, A. (2006). Paraphilic behaviours in a parkinsonian patient with hedonistic homeostatic dysregulation. *International Journal of Neuropsychopharmacology, 9*(6), 767–768. https://doi.org/10.1017/S1461145705006437.

Stein, D. J., Hollander, E., Anthony, D. T., Schneier, F. R., Fallon, B. A., Liebowitz, M. R., & Klein, D. F. (1992). Serotonergic medications for sexual obsessions, sexual addictions, and paraphilias. *Journal of Clinical Psychiatry, 53*(8), 267–271.

Stone, T. H., Winslade, W. J., & Klugman, C. M. (2000). Sex offenders, sentencing laws and pharmaceutical treatment: A prescription for failure. *Behavioral Sciences & the Law, 18*(1), 83–110.

Sturm, V. E., Perry, D. C., Wood, K., Hua, A. Y., Alcantar, O., Datta, S., Rankin, K. P., Rosen, H. J., Miller, B. L., & Kramer, J. H. (2017). Prosocial deficits in behavioral variant frontotemporal dementia relate to reward network atrophy. *Brain and Behavior, 7*(10), e00807. https://doi.org/10.1002/brb3.807.

Tenbergen, G., Wittfoth, M., Frieling, H., Ponseti, J., Walter, M., Walter, H., Beier, K. M., Schiffer, B., & Kruger, T. H. (2015). The neurobiology and psychology of pedophilia: Recent advances and challenges. *Frontiers in Human Neuroscience, 9*, 344. https://doi.org/10.3389/fnhum.2015.00344.

Tewksbury, R., & Lees, M. B. (2007). Perceptions of punishment: How registered sex offenders view registries. *Crime & Delinquency, 53*(3), 380–407.

Thibaut, F., De La Barra, F., Gordon, H., Cosyns, P., Bradford, J. M., & WFSBP Task Force on Sexual Disorders. (2010). The World Federation of Societies of Biological Psychiatry (WFSBP) guidelines for the biological treatment of paraphilias. *The World Journal of Biological Psychiatry, 11*(4), 604–655. https://doi.org/10.3109/15622971003671628.

Tost, H., Vollmert, C., Brassen, S., Schmitt, A., Dressing, H., & Braus, D. F. (2004). Pedophilia: Neuropsychological evidence encouraging a brain network perspective. *Medical Hypotheses, 63*(3), 528–531. https://doi.org/10.1016/j.mehy.2004.03.004.

Tsai, R. M., & Boxer, A. L. (2014). Treatment of frontotemporal dementia. *Current Treatment Options in Neurology, 16*(11), 319. https://doi.org/10.1007/s11940-014-0319-0.

Umbach, R., Berryessa, C. M., & Raine, A. (2015). Brain imaging research on psychopathy: Implications for punishment, prediction, and treatment in youth and adults. *Journal of Criminal Justice, 43*(3), 295–306.

Warren, J. D., Rohrer, J. D., & Rossor, M. N. (2013). Clinical review. Frontotemporal dementia. *BMJ, 347*, f4827. https://doi.org/10.1136/bmj.f4827.

Dr. Cristina Scarpazza is Assistant Professor at the Department of General Psychology, University of Padova. Her research interest lies in psychology and neuroscience, with particular emphasis on early diagnosis of psychiatric disorders, identification of neuroanatomical signature of psychiatric illness, group to individual inferences, and forensic psychiatry (with particular focus on insanity evaluation). She is particularly interested in cognitive biases and their impact in the interpretation of scientific findings. Through her long-standing collaboration with the King's College London, she was actively involved in different projects aiming to improve the translational impact of neuroimaging findings from research to clinical practice.

Dr. Colleen Berryessa is assistant professor at the school of a criminal justice at Rutgers university. In her research, utilizing both qualitative and quantitative methods, she considers how different psychological processes, perceptions, attitudes, and social contexts affect the criminal justice system, particularly related to courts and sentencing. She primarily examines these issues, using both social psychological and socio-legal lenses, in relation to two areas: (1) how these phenomena affect the discretion of criminal justice actors in their responses to offending and decision-making in courts; (2) how these phenomena affect lay views and consideration of courts, sentencing systems, and punishment practices.

Dr. Farah Focquaert is Professor of philosophical anthropology at the Department of Philosophy and Moral Sciences, Ghent University, affiliated with the Bioethics Institute Ghent and Co-director of the International Justice Without Retribution Network. Her research interest lies in the philosophy of free will, responsibility and punishment, and in the field of neuroethics. She is the first editor of the Routledge Handbook of the Philosophy and Science of Punishment (2021).

Three Rationales for a Legal Right to Mental Integrity

Thomas Douglas and Lisa Forsberg

Many states recognize a legal right to bodily integrity, understood as a right against significant, nonconsensual interference with one's body. In this chapter, we offer three rationales for the recognition of an analogous legal right to *mental* integrity.[1]

---

[1]The right to bodily integrity is sometimes explicitly recognized. For example, Article 3(1) of the EU Charter of Fundamental Rights—the right to integrity of the person—states that: 'Everyone has the right to respect for his or her physical and mental integrity'. However, it is more commonly recognized implicitly. In English law, for instance, the right is implicit in the fact that nonconsensual touching of another can incur liability in either or both civil law (battery or assault) or criminal law (assault). A legal right to *mental* integrity could have a similar structure or could be explicitly recognized in a specific civil wrong or a criminal offence; we take no view on this here.

T. Douglas (✉) · L. Forsberg
University of Oxford, Oxford, UK

T. Douglas
Jesus College, Oxford, UK

L. Forsberg
Somerville College, Oxford, UK

S. Ligthart et al. (eds.), *Neurolaw*, Palgrave Studies in Law, Neuroscience, and Human Behavior,
https://doi.org/10.1007/978-3-030-69277-3_8

Introduction

Suppose that an intruder creeps into your bedroom while you are sleeping, pierces your skin with the needle of a syringe, and injects the contents of the syringe into your muscle. And suppose that you knew nothing in advance of his plan to do this.

Clearly, the intruder has wronged you. How has he wronged you? Perhaps he has wronged you by causing you to experience some unpleasant or unwanted state. Perhaps the substance that he has injected will cause you to feel queasy, or lightheaded, or weak. But suppose that all he injected was a tiny amount of sterilized saline. And suppose this has no noticeable effect on you. Still, the intruder seems to have wronged you. How?

Perhaps he has wronged you merely by entering your bedroom without your permission. Perhaps this alone amounts to a trespass on your property or an invasion of your privacy. But this cannot be the whole story, for surely the wrong the intruder perpetrates against you is a greater wrong than the wrong that he would have committed had he entered your bedroom, but without injecting you with anything. His injecting you with a substance seems to have made a moral difference.

One plausible explanation of the difference made by his injecting you would invoke the idea of a right to bodily integrity, understood here as a right against (certain kinds of) significant, nonconsensual bodily interference. By piercing your skin with a needle, he has significantly interfered with your body, and this wrongs you by infringing your right to bodily integrity.

Though it is rarely discussed in detail or fully specified, the right to bodily integrity, as we characterized it above, is often referred to in moral, legal, and political philosophy, albeit not always by that name.[2] This right is often said to be what justifies the moral requirement to obtain consent in relation to medical treatments, organ donation and sex. For instance, Stephen Wilkinson and Eve Garrard (1996, p. 338) suggest

[2]For example, sometimes it is instead referred to as a right against bodily trespass, especially when it is taken to be an implication of self-ownership (see e.g. Thomson, 1990, pp. 205–226; Archard, 2008, pp. 19–34).

that '[o]ne way of explaining the moral significance of organ removal is by appealing to the notion of bodily integrity'. Moreover, the right is often thought to be both uncontroversial and of great importance. Again in the context of organ donation, T. M. Wilkinson (2011, p. 16) states that '[t]he right to bodily integrity… is almost entirely uncontroversial and often considered of great weight'.

Similar thoughts apply at the level of law, where a legal right to bodily integrity is widely recognized. In the context of English law, Baroness Hale held in *R (on the application of Justin West) v The Parole Board* [2002] EWCA Civ 1641, that the right to bodily integrity was 'the most important of civil rights'. In *Re A (Conjoined Twins)* [2001] Fam 147, Walker LJ held that '[e]very human being's right to life carries with it, as an intrinsic part of it, rights of bodily integrity and autonomy—the right to have one's own body whole and intact and (on reaching an age of understanding) to take decisions about one's own body', and in *Collins v Wilcock* [1984] 1 W.L.R. 1172, it was held that '[t]he fundamental principle, plain and incontestable, is that every person's body is inviolate'.[3]

As with the analogous moral requirement, the legal requirement to gain the patient's valid consent to any medical procedure administered to her is often explained by reference to her right to bodily integrity or her right to be free from unlawful touching (and notably consent requirements in respect of medical interventions that do *not* interfere with recipients' bodily integrity are rarely discussed). We see this in medical law textbooks. For example, Emily Jackson begins her chapter on consent to medical treatment as follows: 'One of the first principles of medical law is that patients with capacity must give consent to their medical treatment. *Touching* a person without her consent—however benevolently—is prima facie unlawful' (Jackson, 2019, p. 196, our emphasis). Likewise, Jonathan Herring begins his chapter on consent to medical treatment thus: 'The basic starting point is that a healthcare professional who intentionally or recklessly *touches* a patient without his or her consent is committing a crime (a battery) and a tort (trespass to the person and/or negligence). To be acting lawfully in touching a

[3]For a discussion of the right to bodily integrity in law, see Herring and Wall (2017).

patient, the professional needs a defence' (Herring, 2018, p. 151, our emphasis). And the chapter on consent in *Mason and McCall Smith's Law and Medical Ethics* states that

> Based on the strong moral conviction that everyone has the right of self-determination with regard to his or her body, the common law has long recognised the principle that every person has the right to have his or her bodily integrity protected against invasion by others. Only in certain narrowly defined circumstances may this integrity be compromised without the individual's consent—as where, for example, physical intrusion is involved in the carrying out of lawful arrest. In general, however, a non-consensual touching by another may—subject to the principle *de minimis non curat lex*—give rise to a civil action for damages or, in theory at least, constitute a criminal assault. (Laurie et al., 2019, pp. 65–66)

All of these statements characterize the requirement to obtain the patient's valid consent prior to administering medical interventions to her in terms of respect for or protection of the patient's *bodily integrity*.

Recently, some legal scholars have argued that, just as the law recognizes a right to bodily integrity, so too it should recognize an analogous right to *mental* integrity—a right that we will understand as a right against (certain kinds of) nonconsensual interference with the mind. In their seminal article, 'Crimes Against Minds', Jan Christoph Bublitz and Reinhard Merkel (2014) propose that the law recognize a right to mental self-determination which, they posit, would include a right to 'freedom from mental manipulations' (p. 58) or 'severe [mental] interferences by the state and third parties' (p. 60).[4] As examples of mental interferences, they give, among others, the spiking of drinks in a restaurant with an appetite-enhancing substance (p. 58), use of subliminal imagery by an online store (p. 58), and covert modulation of brain activity using an implanted electrode (pp. 58–59).

[4]The right to mental self-determination would also, they think, include a 'positive dimension', which they characterize as a 'freedom *to* self-determine one's inner realm, e.g. the content of one's thoughts, consciousness or any other mental phenomena' (p. 60, their italics).

Marcello Ienca and Roberto Andorno (2017, p. 5) also argue for the recognition of something like a right to mental integrity, in their case explicitly linking the need for this right to recent developments in the neurosciences. Here, they draw on an analogy with the way in which human rights law responded to the rapid developments in genetic technologies in the last decades of the Twentieth Century. As they note, those developments led to influential declarations concerning the human rights implications of genetic technologies[5]—declarations which effectively recognized new human rights, such as the right not to know one's genetic information.[6] Similar developments will, they suggest, be required in relation to neuroscience: 'the growing sensitivity and availability of neurodevices will require in the coming years the emergence of new rights or at least the further development of traditional rights to specifically address the challenges posed by neuroscience and neurotechnology' (Ienca and Andorno, 2017, p. 8).[7] One new right that they propose is a right that protects 'individuals from the coercive and unconsented use' of emerging neurotechnologies (Ienca & Andorno, 2017, p. 10).[8] This could be understood as a variant of what we are calling the right to mental integrity—one that takes a particular stance on *which* nonconsensual interferences are covered by the right (namely, those that coercively employ neurotechnologies).

Finally, some have argued that what we are calling the right to mental integrity is in fact already strongly protected by international human rights law as one plank of the right to freedom of thought, though it

[5]Universal Declaration on the Human Genome and Human Rights (UDHGHR) 1997, and International Declaration on Human Genetic Data (IDHGD) 2003.

[6]UDHGHR (Art. 5(c)); IDHGD (Art. 10).

[7]See also the Council of Europe's Strategic Action Plan on Human Rights and Technologies in Biomedicine (2020–2025), which, at point 14, explicitly refer to neurotechnology and deep brain stimulation. Available at https://rm.coe.int/strategic-action-plan-final-e/16809c3af1. Accessed 5 June 2020.

[8]Ienca and Andorno understand this right as one aspect of the 'right to cognitive liberty' (with the other aspect being a right to *use* emerging neurotechnologies). They use the term 'right to mental integrity' to refer to a different right: the right to mental health and against mental harm (see esp. p. 18). We prefer to reserve the term 'right to mental integrity' to refer to a right against mental interference since this parallels what we think is the dominant use of the term 'right to bodily integrity'. For other authors who use the term 'right to cognitive liberty' to refer to (something close to) what we call the right to mental integrity, see Sententia (2004), Bublitz (2013), and Bublitz (2015).

has not been adequately developed or enforced. Susie Alegre (2017), for instance, argues that the right to freedom of thought, which is asserted by article 9 of European Convention on Human Rights (ECHR), and article 18 of the International Covenant on Civil and Political Rights, includes a right 'not to have one's thoughts or opinions manipulated' (p. 225), where 'thought' is to be understood broadly and not limited, for example, to only serious or important beliefs (p. 224). Others have argued, more restrictedly, that the existing right to freedom of thought entails rights against 'state indoctrination' by the State or 'brainwashing' (Vermeulen & Roosmalen, 2018, p. 738).[9]

The right to mental integrity has, then, made an appearance in legal scholarship. Thus far, however, the arguments for its recognition remain unclear. Though existing work has *motivated* the claim that we ought to accept such a right—has done much to establish the *prima facie* plausibility of this claim—it falls short of offering a systematic account of the rationales for it. In this chapter, we seek to make some progress towards such a systematic account by delineating and beginning to develop three distinct rationales for the recognition of a legal right to mental integrity: the appeal to intuition, the appeal to justificatory consistency, and the appeal to technological development. In doing so, we will be drawing significantly on the aforementioned work of others—indeed we limit ourselves to considering rationales that are suggested by that work—but we will also be building upon it.

Before proceeding with this task, however, we need to offer a number of qualifications.

First, a crucial distinction: the distinction between *legal* rights and *moral* rights. The abovementioned proponents of the right to mental

[9] European Council's handbook on Article 9 (https://www.echr.coe.int/LibraryDocs/Murdoch2012_EN.pdf), especially p. 18. For other arguments to the effect that article 9 protects the right to mental integrity, see Bublitz (2014) and McCarthy-Jones (2019).

Article 8 ECHR—the right to private and family life—also offers some protection for what we have called the right to mental integrity. The ECtHR held in *Pretty v United Kingdom* that 'the concept of 'private life' is a broad term not susceptible to exhaustive definition', which 'covers the physical and psychological integrity of a person' (*Pretty v United Kingdom Application No 2346/02*, Merits, 29 April 2002). Article 8 should be interpreted in the light of present-day conditions, thus taking into account, inter alia, technological developments and ethical issues to which they may give rise. It seems plausible, then, that article 8 protects at least some aspects of mental integrity in addition to bodily integrity.

integrity appear to think of it as a legal right. For example, in discussing the parallel with genetic rights, Ienca and Andorno make clear that they are thinking of rights that might be created through international declarations, and it is very doubtful that moral rights can be created in this way. Similarly, Bublitz and Merkel defend their proposal in part by arguing that a right to mental integrity (of some kind) is already implicit in the law (in at least some jurisdictions). This would be a strange way to argue for a moral right since the law may be morally mistaken.

In what follows, we will likewise consider only the question whether we ought to recognize a *legal* right to mental integrity (henceforth sometimes an 'LRMI'). Some of the arguments that we give could be re-purposed as arguments for a moral right to mental integrity, but we will not pursue such re-purposing here.

Second, though our focus will be on a legal right, we will be interested in moral, and not legal, rationales for the right. A legal rationale is the sort of rationale that would matter to a court seeking to settle a case. It would establish the LRMI by appealing to existing law. It might, for example, seek to derive the LRMI from some already recognized legal right, as in Alegre's (2017) derivation from the right to freedom of thought, or to show that, as Bublitz and Merkel (2014) suggest, it is pervasively *implicit* in existing law. A *moral* rationale is, by contrast, the sort of rationale that would be of interest to the policymaker given the task of determining whether to recognize an LRMI and placed under no legal obligation to do so, or not to do so. It might, for example, seek to show that the recognition of an LRMI could be supported by plausible moral judgements, principles or theories. In this chapter, we will have nothing to say about legal rationales for the right to mental integrity, but will seek to distinguish and develop three moral rationales.

Third, a limitation on the implications of our discussion. We will, in what follows, primarily be developing—not critiquing—rationales for the LRMI. However, we take no stance on whether these rationales ultimately succeed in justifying the recognition of a LRMI. We are not at all convinced that they do, and everything we say is consistent with there in fact being a decisive case *against* recognizing such a right.

Fourth, a comment on the scope of the LRMI. We acknowledge that there will be immense difficulties in specifying the scope of the right, in part because of difficulties defining the boundaries of the mind, and in part because it is unclear exactly which kinds of nonconsensual interference with the mind would infringe the right to mental integrity. We take it to be plausible that some ways of nonconsensually influencing (and arguably interfering with) the mind will not infringe the right to mental integrity, just as there are ways of nonconsensually influencing a person's body that do not infringe their right to bodily integrity. One reason that some influences on the body fail to infringe the right to bodily integrity is that their effects on the body are not significant enough. If I wave my hand near your arm, causing the hairs on your arm to quiver, I have not infringed your right to bodily integrity, even if I do this without your consent; the effect of the influence is not significant enough. Similarly, there may be mental influences that fail to infringe the right to mental integrity because their mental impact is too insignificant. Another reason that some influences on the body fail to infringe the right to bodily integrity is that they do not employ the required means. If I tell you a disgusting story, causing you to wretch, I do not infringe your right to bodily integrity, even though causing this same bodily reaction though other means—for example, through spiking your drink—would infringe this right. The means of producing the bodily effect matter here. Similarly, there may be mental influences that fail to infringe the right to mental integrity because they do not employ the required means. Giving someone a persuasive argument might cause significant mental changes, but it is doubtful that it would infringe a person's mental integrity, even if done without consent. Exactly how significant an influence has to be to infringe the right to mental integrity, and which means of influence it must employ, are issues that we set aside for future investigation.

With these qualifications in hand, let us turn to the first argument for the recognition of a right to mental integrity: the argument from intuition.

The Appeal to Intuition

Proponents of the LRMI frequently highlight the laxity of existing legal protections against mental interferences and point out the counter-intuitive implications of this laxity. For example, Bublitz and Merkel (2014, p. 51) introduce their discussion of the LRMI as follows:

> Isn't it a bit strange that unpleasant but rather trivial actions like cutting another's hair, inflicting some seconds of minor bodily pain or even firmly touching (without sexual intent) another person may constitute a criminal offense whereas deliberately causing mental suffering often falls squarely out of the purview of the criminal law?

Later, they develop the point thus:

> Suppose [that] neurotools allow us to achieve what has been attempted—and, in the Maoist case, with partial success—interventions into minds changing desires and beliefs without inflicting pain, harming bodily integrity or the need to indoctrinate persons over extended periods of time. Should governments be allowed to resort to such means?—Obviously not. It appears evident that states must be barred from invading the inner sphere of persons, from accessing their thoughts, modulating their emotions or manipulating their personal preferences. At the very least, such measures are in grave need of justification. But then, there must be a right which protects individuals against such interferences. (Bublitz & Merkel, 2014, p. 61)

What is that right? One suggestion—and the suggestion favoured by Bublitz and Merkel—is that it is the right to mental self-determination, of which the right to mental integrity, as we understand it, is one component.

We think that Bublitz and Merkel are here too quick to move from the view that it is 'obvious' and 'evident' that states should be prohibited from doing certain things to the claim that their doing those things must violate some right. It is possible to explain why states ought to be prevented from 'invading the inner sphere of persons' without appealing to (either moral or legal) rights. Perhaps states ought to be prevented from doing this simply because their doing so will typically cause more harm than good.

Still, we think this passage does suggest an argument in favour of recognizing an LRMI. According to this argument, we ought to recognize a legal right to mental integrity because (a) widely held moral intuitions suggest that there is a distinctive duty not to interfere with others' minds (that is, a *prima facie* moral duty that is distinct from the duty not to interfere with others' bodies), and (b) it would be desirable, or at least permissible, to enforce this distinctive duty through recognizing a legal right to mental integrity.

Claim (b) depends on general considerations regarding the purpose and effectiveness of legal rights that we cannot explore here. We simply take it for granted. Instead, we will focus on (a).

Which intuitions support a distinctive duty not to interfere with the minds of others? We believe that two sets of intuitions are relevant here.

First, there are intuitions to the effect that interventions that interfere with both the body and the mind often seem more seriously wrong, morally, than comparably physically invasive interventions that do not interfere with the mind. Consider the following case:

> Thinking that one of her regular customers looks a little down, a well-meaning but paternalistic barista surreptitiously slips a newly developed mild, short- and fast-acting anti-depressant into his morning coffee, with the result that the customer's mood is somewhat lifted for a few hours.

Call the intervention in this case *Anti-depressant*, and compare it with the following intervention, which we will call *Anti-asthmatic*:

> Thinking that one of her regular customers looks a little wheezy, a well-meaning but paternalistic barista surreptitiously slips a mild, short- and fast-acting anti-asthmatic medication into his morning coffee, with the result that the customer breathes somewhat more easily for a few hours.

It seems to us that *Anti-depressant* is, *prima facie,* more seriously wrong, or wrong in a different way than *Anti-asthmatic*, even though the two interventions seem similar with respect to the nature and degree of bodily interference that they involve. One plausible way to explain the difference, we think, would be to invoke a duty not to interfere with others' minds. While both *Anti-depressant* and *Anti-asthmatic* interfere with your body, and in similar ways and to similar degrees, only *Anti-depressant* interferes with your mind. Thus, *Anti-depressant* infringes an additional duty, and so is more seriously wrong.

A second cluster of intuitions that support a distinctive duty not to interfere with others' minds are intuitions concerning certain physically non-invasive forms of mental interference; interventions that we would commonly refer to as 'brainwashing'. Consider, for example, the possibility that someone might hypnotize you against your will, or seek to alter your desires through subliminal imagery, or subject you to a some kind of aversion therapy in which authority figures subject you to distressing images whenever you perform some undesired behaviour. It is interventions of this sort that Bublitz and Merkel (2014, p. 61) presumably have in mind when they refer to the partial Maoist 'success' in 'changing desires and beliefs'.

It seems intuitively clear that such interventions are typically wrong. Yet we clearly cannot explain this by adverting to bodily interference, since though such forms of brainwashing must induce bodily changes—they could not otherwise affect the mind—they do not plausibly violate any duty by virtue of their bodily effects. A distinctive duty not to interfere with the *minds* of others could, however, explain the wrongness of brainwashing and recognizing an LRMI could help to enforce this duty.

The Appeal to Justificatory Consistency

A second point often emphasized by proponents of a right to mental integrity is that standard theoretical justifications for the right to bodily integrity appear also to support a right to mental integrity. Consider the following from Bublitz and Merkel (2014, p. 62):

> In the wake of Locke, libertarians believe that persons have property rights in their body; persons literally own (the physical part of) themselves. Ownership discussions focus on the relation of persons to their

> bodies, their liberty (e.g. vis-a-vis slavery) and the fruits of their labor. But what is even more constitutive of a subject than her body is her mind. So, whoever grants self-ownership of persons over their bodies has a compelling reason to concede self- ownership over minds.[10]

The suggestion here, we take it, is that, if we are to recognize a legal right to bodily integrity, then we ought, on pain of inconsistency, to recognize at least a defeasible case for a legal right to mental integrity.[11] We ought to do this because the theoretical considerations that justify the right to bodily integrity also provide (defeasible) support to the right to mental integrity. Call this the argument from justificatory consistency.

Whether the appeal to justificatory consistency is compelling will, of course, depend on what considerations justify the right to bodily integrity. Bublitz and Merkel suggest one candidate—self-ownership—but there are others. A full development of the argument would need to survey all plausible justifications and consider whether each supports also a right to mental integrity. We cannot pursue this approach here, but let us briefly introduce some of the most frequently mentioned justifications. These fall into broadly two categories. First, there are *rights-based* justifications; justifications that seek to derive the right to bodily integrity from some more fundamental right. Second, there are *interest-based* justifications; justifications according to which the right to bodily integrity is justified by its role in protecting some interest.

Consider first rights-based justifications. These typically appeal to one of two more fundamental rights: property rights over the self—rights of self-ownership—normally understood as analogous to property rights

---

[10]English law traditionally took the view that there are no property rights in human bodies (see e.g. *R v Bentham* [2005] 1 WLR 1057), with the exception of cases in which the lawful exercise of work and skill has been applied to it (*Doodeward v Spence* [1908] 6 CLR 496; *R v Kelly* [1999] 2 WLR 384). In *Yearworth v North Bristol NHS Trust* [2009] 3 WLR 118, the Court of Appeal held that a property right extended to one's sperm. The Human Fertilisation and Embryology Act 1990 (HEFA) as amended by HEFA 2008 regulates the storage and use of human reproductive materials by consent requirements, rather than as property, and such consent requirements provide limited guidance when conflicts over ownership arise (*Evans and others v Amicus Healthcare* [2003] EWHC 261). Similarly, the Human Tissue Act 2004 regulates the removal, storage, use and disposal of human body parts, organs and tissue by consent, by without treating human materials as property.

[11]A defeasible case is a case that has some normative force, but is not necessarily decisive; it can be defeated by countervailing considerations.

over external property (e.g. Thomson, 1990), and rights to personal sovereignty—understood on analogy with the rights of states over their territory (e.g. Archard, 2008; Ripstein, 2006). Both types of rights attach to the self or person (we take the two to be equivalent, and henceforth use the term 'self'), and both are normally taken to include or imply rights against interference with the self. These rights against interference with the self are in turn thought to imply rights against interference with the body since the body either is, is part of, or is closely connected to, the self.

Though discussions of self-ownership and personal sovereignty more frequently draw out implications for the body than for the mind,[12] it seems clear that appeals to self-ownership or personal sovereignty will also support rights over the mind, since the mind clearly also either is, is part of, or is closely connected to, the self.[13] Indeed, most currently dominant accounts of the self give the mind a more central role than the body in the self. On psychological accounts, for instance, the self is, or resides wholly in, the mind, with the body being merely a contingent receptacle for the self. So, we might think that considerations of self-ownership and personal sovereignty in fact provide stronger support to a moral right to mental integrity than to a moral right to bodily integrity.

Consider next interest-based justifications. These justify the right to bodily integrity by reference to its role in protecting some interest of the right-holder. The interest most commonly invoked is the interest in autonomy, which is frequently analysed as an interest in controlling one's life, and/or in living one's life free from the control or domination of others. The thought is that the right to bodily integrity serves to safeguard our autonomy (e.g. Mill, 1859; Feinberg, 1986).

Note that the claim here need not be that *every* infringement of the right to bodily integrity diminishes a person's autonomy. Rather, the thought may be that, since infringements of the right to bodily integrity

[12]Though for rare explicit acknowledgments that rights over the self will imply rights over the mind, see, for example, Mill (1859, p. 11), who holds that '[o]ver himself, over his own body and mind, the individual is sovereign', and Lippert-Rasmussen (2018, p. 142), who characterises self-ownership as 'moral ownership of himself or herself, that is, his or her body and mind'.

[13]Bublitz and Merkel (2014, esp. 62, 73) make this same point.

tend to diminish autonomy, recognizing a right to bodily integrity is one way (and perhaps part of the best way) to protect autonomy.

Again, it seems clear that a parallel justification would provide defeasible support to a right to mental integrity. After all, interferences with the mind can be just as threatening to autonomy as interferences with the body.

Consider the possibility of nonconsensual hypnosis, mentioned above. Nonconsensual hypnosis is a paradigmatic example both of loss of control over one's life, and subjugation to the control of another. It very plausibly produces a serious loss of autonomy on whichever of the dominant approaches to autonomy one adopts. A right to mental integrity would protect against such interferences.

The Appeal to Technological Development

A third rationale for the LRMI is suggested by the frequent reference, by proponents of the right, to recent and likely future neurotechnological developments. These developments play an especially prominent role in Ienca and Andorno's work. Following a survey of recent advances in neuroscience, they claim (2017, p. 5) that

> if in the past decades neurotechnology has unlocked the human brain and made it readable under scientific lenses, the upcoming decades will see neurotechnology becoming pervasive and embedded in numerous aspects of our lives and increasingly effective in modulating the neural correlates of our psychology and behavior.

This, they suggest (p. 2) creates a possible need for new legal rights:

> the possibilities opened up by neurotechnological developments and their application to various aspects of human life will force a reconceptualization of certain human rights, or even the creation of new rights.

And, on their view, one new right that might need to be created is what we are calling the legal right to mental integrity.

Bublitz and Merkel (2014, p. 65) also emphasize the relevance of technological developments, claiming that the law must erect 'normative boundaries' around the mind now that 'neurotechnologies promise to enable us to surmount the natural boundaries of the mind (the skull) and to modulate the inward workings of the mind'.

These references to neurotechnological developments are, we think, best understood as responses to one or more of a range of potential objections to recognizing an LRMI. These objections hold that, even if there is a sense in which the mind deserves the protection of an LRMI—say, because there is a distinctive duty not to interfere with others' minds—providing such legal protection is unnecessary or undesirable. In what follows, we survey these objections, in each case describing how an appeal to technological developments might undermine the objection.

The most straightforward reason to think that an LRMI would be unnecessary or undesirable, and the one that Bublitz and Merkel and Ienca and Andorno are most concerned to rebut, holds that recognizing an LRMI is unnecessary because the mind is in any case insusceptible to the kinds of interference that would infringe the right. Call this the *insusceptibility objection.*

Both Bublitz and Merkel and Ienca and Andorno acknowledge that, historically, the mind has indeed been regarded as insusceptible to mental interference, or at least, to mental interference of the kinds that seem most morally troubling: *irresistible* interference, or what we might call 'mind control'. Bublitz and Merkel (2014, p. 61) suggest that the right to mental integrity 'has never been considered more thoroughly because, traditionally, the mind has not been conceived as an entity vulnerable to external intrusions and hence in need of legal protection'.[14] Moreover, they concede that at one point this way of thinking may have been justified; in the 1940s, 'there may have been good reasons to emphatically believe in the untouchable absoluteness of freedom of the mind' and 'the factual invincibility of the mental realm' (p. 65). Ienca and

---

[14] Bublitz and Merkel also cite evidence that delegates involved in drafting the Universal Declaration on Human Rights subscribed to this view. For example, one is reported to have held that 'It would be unnecessary to proclaim freedom of [the inner sphere] if it were never to be given an outward expression as the inner is beyond any access' (Bublitz & Merkel, 2014, p. 64, citing Hammer, 2001, p. 34).

Andorno (2017, 1) go further by actually endorsing the past invulnerability of the mind to external control: 'While the body can easily be subject to domination and control by others, our mind, along with our thoughts, beliefs and convictions, [have until recently been] to a large extent beyond external constraint'. Both sets of authors, however, suggest that, if the mind was ever insusceptible to irresistible interference, it is no longer so; the insusceptibility objection to recognizing an LRMI no longer holds, and the appeal to technological developments explains why.

Perhaps, however, the objection can be reintroduced in a more plausible form. It might be held that the LRMI is unnecessary not because the mind is insusceptible to interference, but because almost all forms of mental interference that might plausibly infringe the right can already be legally regulated in other more straightforward ways. For example, it might be held that the vast majority of interventions that would infringe the LRMI would also infringe the—already established—legal right to *bodily* integrity and could be satisfactorily regulated on that basis. True, some extreme forms of brainwashing, such as nonconsensual hypnosis, would presumably infringe a legal right to mental integrity without infringing the right to bodily integrity. But these interventions are arguably vanishingly rare. *Most* interferences with mental integrity involve the administration of drugs or other neurotechnologies. These interventions are somewhat physically invasive, and so can perfectly well be regulated on the basis that they infringe the right to bodily integrity. Call this the *existing protection objection.*

The cogency of the existing protection objection will clearly depend on which forms of mental interference, exactly, would infringe the right to mental integrity. Nonconsensual neurointerventions and the various forms of brainwashing are obvious candidates, but there is much grey area. We could legitimately wonder, for example, whether many so-called nudges might infringe the right.

Consider the famous cafeteria nudge, in which cafeteria staff place healthier foods at eye level in a cafeteria, knowing that they will then appear more salient to, and be more likely to be chosen by, cafeteria customers. Or consider the practice of serving food on smaller plates, to make a given serving size appear larger. These practices clearly involve attempts to intentionally influence a person's preferences. It would not, we think, be too much of a stretch to refer to them as instances of mental interference, and it thus seems possible that they would infringe a right to mental integrity, should we possess such a right.

Consider alternatively the myriad practices sometimes employed by the designers of computer games and online services for the purposes of promoting 'customer engagement'. We might think, in this connection, of the use of randomized rewards to promote addiction to computer games, or the use of bottomless newsfeeds to keep users of social media platforms online. Again, these practices might aptly be characterized as mental interference and again, and it thus seems possible that they might infringe a right to mental integrity.

If the right to mental integrity is understood very broadly, so as to include possibilities such as those we have just mentioned, then it seems clear that the existing protection objection to recognizing an LRMI fails: on a broad construal, the right to mental integrity will cover many interventions that are *not* physically invasive and so cannot be regulated as infringements of bodily integrity.

But suppose that the right to mental integrity should instead be understood narrowly. Suppose that it covers only those interventions that obviously involve problematic forms of mental interference. Could we then—as the existing protection objection maintains—get by with only a right to bodily integrity?

It is not clear that we could, and this is where the appeal to technological development again enters the scene. The claim might be made that we are likely, in the near future, to have at our disposal many means of mental interference that (i) would obviously infringe a right to mental integrity and (ii) cannot be adequately regulated under a right to bodily integrity.[15]

Indeed, some such technologies arguably already exist. Consider transcranial direct current stimulation (tDCS) and transcranial magnetic stimulation (TMS)—interventions that act on the mind by subjecting the brain to a small electric current or magnetic field. These forms of brain stimulation have been shown to be capable of modulating various aspects of mental functioning including mood, working memory,

[15]Bublitz and Merkel, 'Crimes Against Minds' (2014, esp. 60) make a similar point.

cravings for addictive substances, and numerical processing ability.[16] Though they typically involve devices—either electrodes or magnets—being placed on the scalp, it is plausible, at least for TMS, that the procedure could be performed without any touching—with the magnets held slightly above the scalp.

Consider this intervention, which we will call *Nonconsensual TMS*:

> Your housemate, a budding neuroscientist, notices that you seem to have had a sore leg for the last few days, since completing a half-marathon. To help reduce the pain, she sneaks into your room one evening and, without your prior knowledge, subjects you to transcranial magnetic stimulation (TMS). This involves applying a magnetic field to the parts of the brain responsible for the sensation of pain using magnets placed just above the scalp. It does not involve any physical touching. The procedure succeeds in diminishing the pain that you feel over the coming days, and does not at all affect the underlying cause of the pain.

It seems clear that, in implementing *Nonconsensual TMS*, your housemate acts wrongly. It is plausible that the law ought to protect you against this intervention. However, it also seems doubtful that *Nonconsensual TMS* could be adequately regulated under the head of bodily integrity, given that it involves no touching.

Similar thoughts apply to the nascent technology of optogenetics, which involves the use of light to modulate the activity of (typically genetically modified) neurones. This intervention has been shown to be capable of modulating fear (e.g. Dias et al., 2013) and erasing and re-inserting memories (e.g. Nabavi et al., 2014). In cases where superficial brain areas are targeted, the light can be administered through the skull, without the need for internal light sources. Again, it seems clear that nonconsensual uses of this technology to significantly alter a person's mental states would typically be wrong. It also seems plausible that they could not be adequately regulated as infringements of the right to bodily integrity, given that shining a light through a person's skull need involve no touching.

---

[16] For a review of the effects of tDCS and other forms of non-invasive brain stimulation, see Polanía et al. (2018).

We have outlined how technological developments might be invoked to diffuse two possible objections to recognizing an LRMI: the *insusceptibility objection* and the *existing protection objection.* Let us now turn to consider a third objection. It might be held that recognizing an LRMI would be undesirable because, even if the right could be precisely defined, in practice it would be too difficult to identify infringements of it, since it is difficult to identify changes to a person's mental states. As Bublitz and Merkel (2014, p. 52) note, difficulties in identifying mental changes have led to a general reluctance to legally protect the mind: "Mental states, thoughts, feelings, behavioral dispositions hidden from view in the "inner citadel" of the individual's consciousness are regarded as intangible, evanescent, too elusive for the law to handle".

Again, however, this objection may be undermined by technological developments—for example, in neuroimaging—which could allow for more accurate identification of mental alterations. As Bublitz and Merkel (2014, p. 53) write,

> what especially brings the venerable issue of mental harms back on the table of legal theory is neuroscience, promising to reveal subjective states as grounded in objective facts, i.e. in events observable from the third-person perspective. When mental states lose their empirical intractability, the legal disregard for the mind loses its plausibility.

In this connection, it is important to note that the law does already seek to regulate some effects on the mind. For example, English criminal law accepts that *actual bodily harm* (ABH) for the purposes of the Offences Against the Person Act 1861 comprises psychiatric or psychological harm in addition to harm directly inflicted on the body. The Crown Prosecution Service advises that '[p]sychological harm that involves more than mere emotions such as fear, distress or panic can amount to ABH', but that 'psychological injury not amounting to recognizable psychiatric illness does not fall within the ambit of bodily harm for the purposes of the 1861 Act'.[17] In order for psychiatric or psychological injury to

[17]Crown Prosecution Service, *Offences against the Person, incorporating the Charging Standard*, available at https://www.cps.gov.uk/legal-guidance/offences-against-person-incorporating-charging-standard, updated 6 January 2020. *R v Chan-Fook* [1993] EWCA Crim 1; *R v Ireland*

amount to ABH, it must be supported by appropriate medical expert evidence.

An LRMI would, we suppose, cover a much broader range of mental alterations than existing protections against ABH. For example, just as the right to *bodily* integrity protects against even non-harmful forms of bodily interference, we might expect a right to mental integrity to protect against non-harmful forms of mental interference. Many such interferences would presumably involve much more subtle mental alterations than those which constitute psychiatric or psychological injury—alterations for which it would historically have been difficult to provide reliable, objective evidence. However, new technologies may help to provide reliable, objective evidence of a broader range of different kinds of mental alteration, including many that would like beyond the scope of ABH.[18]

Concluding Thoughts

We have identified and outlined three distinct rationales for recognizing a legal right to mental integrity, drawing on comments previously made by others to motivate the recognition of this right: the appeal to intuition, the appeal to justificatory consistency, and the appeal to technological development.

Each of these rationales is open to question. For example, one could attempt to rebut the appeal to intuition by maintaining that there are better ways to legally enforce the distinctive duty not to interfere with others' minds than by recognizing an LRMI, one could attempt to rebut the appeal to justificatory consistency by denying that we ought to recognize a right to bodily integrity, and one could attempt to rebut all three appeals by maintaining that enforcing an LRMI would—even given technological developments—be too costly. Nevertheless, we think that each of these candidate rationales has some plausibility and warrants

[1998] CA 147; *Director of Public Prosecutions v Smith* [2006] EWCA 94; *R v D* [2006] EWCA Crim 1139.

[18]For criticisms of the law relating to psychiatric injury, see e.g. Teff (2009) and Ahuja (2015).

further scrutiny. We hope that by outlining and distinguishing them, we will encourage such scrutiny.

References

Ahuja, J. (2015). Liability for psychological and psychiatric harm: The road to recovery. *Medical Law Review, 23*(1), 27–52.

Alegre, S. (2017). Rethinking freedom of thought for the 21st century. *European Human Rights Law Review, 3,* 221–233.

Archard, D. (2008). Informed consent: Autonomy and self-ownership. *Journal of Applied Philosophy, 25*(1), 19–34. https://doi.org/10.1111/j.1468-5930.2008.00394.x.

Bublitz, J. C. (2013). My mind is mine!? Cognitive liberty as a legal concept. In E. Hildt & A. Franke (Eds.), *Cognitive enhancement. Trends in augmentation of human performance* (Vol. 1). Dordrecht: Springer.

Bublitz, C. (2014). Freedom of thought in the Age of neuroscience. *Archiv Für Rechts- Und Sozialphilosophie, 100,* 1–25.

Bublitz, J. C. (2015). Cognitive liberty or the international human right to freedom of thought. In J. Clausen & N. Levy (Eds.), *Handbook of neuroethics.* Dordrecht: Springer.

Bublitz, J. C., & Merkel, R. (2014). Crimes against minds: On mental manipulations, harms and a human right to mental self-determination. *Criminal Law and Philosophy, 8*(1), 51–77. https://doi.org/10.1007/s11572-012-9172-y.

Dias, B. G., et al. (2013). Towards new approaches to disorders of fear and anxiety. *Current Opinion in Neurobiology, 23*(3), 346–352.

Feinberg, J. (1986). *Harm to self*. Oxford: Oxford University Press.

Hammer, L. (2001). *The international human right to freedom of conscience.* Dartmouth: Ashgate.

Herring, J. (2018). *Medical law and ethics* (7th ed.). Oxford: Oxford University Press.

Herring, J., & Wall, J. (2017). The nature and significance of the right to bodily integrity. *Cambridge Law Journal, 76*(3), 566–588. https://doi.org/10.1017/S0008197317000605.

Ienca, M., & Andorno, R. (2017). Towards new human rights in the age of neuroscience and neurotechnology. *Life Sciences, Society and Policy, 13,* 5. https://doi.org/10.1186/s40504-017-0050-1.

Jackson, E. (2019). *Medical law: Cases, text and material* (5th ed.). Oxford: Oxford University Press.

Laurie, G., Harmon, S., & Dove, E. (2019). *Mason and McCall Smith's Law and medical ethics*. Oxford: Oxford University Press.

Lippert-Rasmussen, K. (2018). The self-ownership trilemma, extended minds, and neurointerventions. In D. Birks & T. Douglas (Eds.), *Treatment for crime: Philosophical essays on neurointerventions in criminal justice*. Oxford: Oxford University Press.

McCarthy-Jones, S. (2019). The autonomous mind: The right to freedom of thought in the twenty-first century. *Frontiers in Artificial Intelligence, 2*(19), 1–17.

Mill, J. S. (1975) [1859]. *On liberty* (D. Spitz, Ed.). Toronto: W. W. Norton.

Nabavi, S., et al. (2014). Engineering a memory with LTD and LTP. *Nature, 511*(7509), 348–352.

Polanía, R., Nitsche, M. A., & Ruff, C. C. (2018). Studying and modifying brain function with non-invasive brain stimulation. *Nature Neuroscience, 21,* 174–187.

Ripstein, A. (2006). Beyond the harm principle. *Philosophy & Public Affairs, 34*(3), 215–245.

Sententia, W. (2004). Neuroethical considerations: Cognitive liberty and converging technologies for improving human cognition. *Annals of the New York Academy of Sciences, 1013*(1), 221–228.

Teff, H. (2009). *Causing psychiatric and emotional harm: Reshaping the boundaries of legal liability*. Oxford: Hart Publishing.

Thomson, J. J. (1990). *The realm of rights*. Cambridge, MA: Harvard University Press.

Vermeulen, B., & Roosmalen, M. (2018). Freedom of thought, conscience and religion. In P. van Dijk, F. van Hoof, A. van Rijn, & L. Zwaak (Eds.), *Theory and practice of the European convention on human rights* (5th ed.). Cambridge: Intersensia.

Wilkinson, S., & Garrard, E. (1996). Bodily integrity and the sale of human organs. *Journal of Medical Ethics, 22*(6), 334–339.

Wilkinson, T. M. (2011). *Ethics and the acquisition of organs*. Oxford: Oxford University Press.

Thomas Douglas is Professor of Applied Philosophy at the Oxford Uehiro Centre for Practical Ethics, where he is Director of Research and Development. He is also a Senior Research Fellow at Jesus College, Editor of the Journal of Practical Ethics, and Principal Investigator on the project 'Protecting Minds: The Right to Mental Integrity and the Ethics of Arational Influence', funded by a Consolidator Award from the European Research Council. His research lies mainly in practical and normative ethics and currently focuses on the ethics of predicting and influencing behaviour.

Dr. Lisa Forsberg is a British Academy Postdoctoral Fellow in the Faculty of Law, and (in Philosophy) at Somerville College and the Oxford Uehiro Centre for Practical Ethics. Her main research interests lie in normative and practical ethics, and in the philosophy of medical and criminal law. Her postdoctoral project, 'Changing One's Mind: Neurointerventions, Autonomy, and the Law on Consent', is on medical consent and examines the extent to which English law on consent sufficiently protects morally salient patient interests.

Neurointerventions and Crime Prevention: On Ideal and Non-ideal Considerations

Jesper Ryberg

Introduction

Is it morally acceptable (or required) to use neurointerventions on criminal offenders as an instrument to prevent future engagement in criminal activity? This question is not one that allows for an immediate and simple answer. On the contrary, as with many other ethical challenges, an answer cannot be given without engaging in initial conceptual clarification.

For instance, what kind of neurointervention are we considering? It is today a well-known fact that the term "neurointerventions" covers a wide range of treatment options varying from drugs, over non-invasive techniques (e.g. transcranial magnetic stimulation (TMS) and transcranial direct current stimulation (tDCS)), to invasive techniques (e.g. deep

J. Ryberg (✉)
Department of Philosophy, Roskilde University, Roskilde, Denmark
e-mail: ryberg@ruc.dk

S. Ligthart et al. (eds.), *Neurolaw*, Palgrave Studies in Law, Neuroscience, and Human Behavior,
https://doi.org/10.1007/978-3-030-69277-3_9

brain stimulation (DBS) or other kinds of brain surgery). These treatment techniques may of course affect those who are being treated very differently. Moreover, the question of whether it is acceptable to "use" such interventions is clearly ambiguous. A neurointervention can either be used by merely offering it to an offender, by administering it in a way that is designed to incentivize the offender to accept the treatment or by subjecting the offender to outright compulsory treatment.[1] Needless to say, these ways of using a neurointervention may have very different consequences for the person upon whom the treatment is administered. When these two aspects of a clarification are added to the fact that a crime preventive effect can in itself vary significantly in scope, the initial question will cover both of the following polar extremes. On the one hand, one can ask whether it would be acceptable to offer an offender a pill with very few side effects but with a comprehensive crime preventive effect. On the other, the question comprises the possibility of compulsorily administering a brain-surgical technique with severe effects on the life of the offender and with an uncertain, marginal crime preventive effect. While the first possibility is hard to object to—if the offender is free to choose and fully informed—the latter is clearly morally unacceptable. Thus, much of the academic discussion that has in recent years commenced on crime preventive use of neurointerventions has been concerned with the challenge as to where the ethical lines more precisely should be drawn within the framework of these diametrically different explications of the question.

The purpose of this chapter is not to contribute to this discussion by considering the ethical legitimacy of different ways of administering different types of neurointervention. Rather, the point in the following will be to direct attention to yet another distinction which is crucial to keep in mind in considerations of how the initial question should be interpreted. More precisely, the question may be interpreted as asking whether it can *ever* be justified to use neurointerventions in a particular way to prevent recidivism; or, alternatively, whether it would be justified to use neurointerventions within the criminal justice context that

[1] For a comprehensive ethical discussion of these different ways of administering neurointerventions, see Ryberg (2020).

currently exists (or will exist in the near future).[2] As we will see, these two ways of understanding the question—to wit, whether it would in principle or practice be acceptable to use neurointerventions for crime preventive purposes—may lead to very different answers (see also Ryberg, 2020). In order to clarify and buttress the significance of this distinction, the chapter will proceed as follows.

In section "Ideal Penal Theory and Non-ideal Penal Practice", it will be shown how criminal justice practice diverges significantly from what is desirable from a penal theoretical point of view. The developments in the US will be used as the example. Some of the aspects of this development are highlighted, and it is suggested that they may have implications for the way in which neurointerventions might be used if such treatment options were to be adopted as a method of preventing offenders from returning to crime. In section "Current Use of Neurointerventions", it is briefly suggested that some of the ways one might fear that neurointerventions would be misused under the actual non-ideal circumstances cannot be rejected by reference to the ways in which some neurointerventions are already being used in parts of criminal justice practice. Section "The Significance of the Ideal/Non-ideal Distinction for Ethical Theorizing" elaborates on the significance of distinguishing between ideal and non-ideal theorizing, by showing how a failure to comprehend and account for this distinction involves a risk of leading the ethical discussion of neurointerventions astray. Finally, section "Conclusion" summarizes and concludes.

Thus, the overall point of this chapter is to emphasize the distinction between ideal and non-ideal considerations which, in my view, has not yet received sufficient attention but which is crucial to acknowledge if one subscribes to the generally shared view that the goal of engaging in neuroethical considerations is not mere theoretical puzzle-solving but is to provide genuine action-guidance with regard to whether and how neurointerventions should be used in criminal justice practice.

---

[2] The distinction between ideal and non-ideal theorizing, to which I will refer in the following, has been used in slightly different ways in modern ethics and political philosophy. For a fine overview, see Valentini (2012).

Ideal Penal Theory and Non-ideal Penal Practice

Administering neurointerventions within the criminal justice system—however precisely this is done—is tantamount to using such treatment in a framework which in itself is subject to comprehensive ethical criticism. The ethics of punishment is almost never regarded as providing a defence of the existing penal order, but rather as considerations establishing that criminal sanctions are very often not used in a way that is ethically justified. In fact, the discrepancy between actual penal practice and what would constitute the ideal use of punishment is often regarded as significant. A repeatedly highlighted example is the use of punishment in the US.[3]

The story of the development in the use of punishment in the US over the last four to five decades has often been told. Until the 1970s, rates of imprisonment had been fairly stable for most of the twentieth century (Tonry, 2009). However, in the following years, this picture changed dramatically. In the 1970s, the imprisonment rate started increasing. Even though crime rates declined in the early 1990s, prison growth accelerated and continued to rise until the end of the first millennium (reaching a peak at just above 750 inmates per 100.000 population). Over the last couple of years, there have been some fluctuations in the imprisonment rate. However, the overall picture is that the rate has quintupled over a few decades. Even though there is no general agreement on what has initiated and fuelled this remarkable development, there is widespread adherence to the view that something has gone entirely wrong; that is, that a situation has resulted in which many offenders are being punished *too* severely.

The latter claim is of course a normative judgement. However, despite the fact that there is no general agreement among penal theorists with regard to what constitutes the most plausible ethical theory of punishment, the view that penal practice deviates significantly from what is ethically desirable is one to which all theorists subscribe. For instance,

[3]For a more comprehensive presentation and discussion of the issues in this section, see also Ryberg (2020, chap. 7).

retributivist-minded theorists have underlined that it is a mistake to think of the desert-based theory of punishment as a derivative of a "throw away the key" approach, and that if this theory was followed consistently, one would punish less and in more decent ways than one actually does.[4] In fact, a leading retributivist such as Andreas von Hirsch has even suggested that his theory should be seen as a way of restricting punishment and that terms of imprisonment, even for the most serious crimes, should seldom exceed five years (von Hirsch, 1993; von Hirsch & Ashworth, 2005).

Along the same lines—though from a different perspective—consequentialist-minded theorists have repeatedly underlined that studies of crime prevention have consistently shown that there is no preventive gain by locking offenders up for longer periods. For instance, as part of his discussion of the effects of punishment in the US penal system, Michael Tonry has summarized his recent review of studies on crime prevention in the following way:

> In 2017, it is not controversial to assert that the crime prevention effects of mass incarceration have been much less than many people supposed or hoped, that there is little or no reason to believe that harsher punishments have greater deterrent effects than milder punishments, that incapacitating people by locking them up for lengthy periods is an ineffective crime prevention strategy, or that the experience of imprisonment makes offenders more not less, likely to commit crime later in their lives.

and, consequently, that:

> The implications of the literatures on deterrence and incapacitation are straightforward: few convicted offenders should be sent to prison and for shorter time. (Tonry, 2016, pp. 453 and 459)

Thus, both retributivists and consequentialists agree that something has gone entirely wrong. When it comes to the question as to how this development should be explained—that is, how one has ended up in

[4] See, for instance, Singer (1979) or Murphy (1979). For an overview, see also Ryberg (2004, 2020).

a situation that has been described as a "tragedy" and "national embarrassment" (Tonry, 2016, p. 441; Husak, 2019)—there is less agreement among penal theorists, except with regard to the fact that many factors have contributed.[5] In the following, I will not try to outline the contours of competing overall explanations. Rather, the point here is to illustrate that, if one imagines that crime preventive use of neurointervention was to be implemented within the framework of such a non-ideal criminal justice practice, then there would also be reasons to fear that such use might diverge from what is ethically desirable. Here are three ways in which this could be the case.

Overuse of neurointerventions. As initially noted, neurointerventions can be administered on offenders in different ways. For instance, such treatments can be offered as a condition for parole or release from prison. They can also be forced upon offenders. Suppose that neurointerventions were administered in either of these ways. One aspect of an ethical assessment of such ways of administering neurointerventions would of course be whether they are used in a way warranted by consequentialist crime preventive considerations. The answer to this depends both upon whether there will be a preventive effect and on how the treatment may adversely affect the person upon whom the treatment is imposed. Even though there may be significant differences between different types of neurointerventions, it is a well-known principle of pharmacology that treatments have side effects. And the imposition of a neurointervention may also have various deleterious psychological effects. But this means that if neurointerventions were to be used in some cases where there is no crime preventive gain, then such treatment would be wrong. But are there any reasons to fear that neurointerventions might be inflicted on offenders if they do not have a beneficial effect in terms of crime prevention? Under ideal conditions, the answer would obviously be in the negative. After all, the whole point of the sort of neurointerventions we are considering here is that they should serve a crime preventive function. If this function does not exist, the use of such treatment would amount to an unnecessary imposition of hardship or suffering on offenders.

[5] For discussions and different views of the various factors that have contributed to this development, see, e.g., Garland (2001), Lacey (2008), Pfaff (2017), and Tonry (2004, 2009).

However, if the question is posed under non-ideal circumstances, the answer may turn out very differently.

In order not to administer pointless treatment on offenders, one would have to draw on assessments of the risk that an offender might fall back into crime. The first thing that should be noted, therefore, is that in real-life risk assessment tools are far from accurate. For instance, a meta-analysis on the prediction of violence concluded that, on average, positive predictions were correct 42% of the time (Fazel et al., 2012).[6] Furthermore, there are problems when it comes to the use of group predictions in individual cases. For instance, Seena Fazel et al. have underlined that "risk assessment tools in their current form can only be used to roughly classify individuals at the group level, and not to safely determine criminal prognosis in an individual case" (Fazel et al., 2012, p. 5). However, more importantly, even if one imagines that the accuracy of instruments of crime prediction was significantly improved, it cannot be taken for granted that real-life decisions on the use of neurointerventions would be based on such predictions. One of the many mechanisms which criminologists and other theorists have suggested may have contributed to the current problem of over-punishment is the so-called "false-positive problem" (see, e.g., Pfaff, 2017). In relation to the use of risk-based sentencing, there exist two types of mistake. One mistake would be not to send someone to prison—or to fail to keep an offender in prison for a longer period—who commits a crime that could have been prevented had he or she been behind bars. This is a false-negative result. Another mistake would consist in incarcerating someone who does not constitute a genuine risk, that is, who would not have committed a crime had he or she been free. This is a false-positive error. Now, the false-positive problem consists in the fact that these two types of error may be reacted to in very different ways. John F. Pfaff has appositely phrased the problem in the following way:

> The costs of a false negative are immediate and salient. Those of a false positive are nearly invisible and abstract. In the case of a false negative, there is an identifiable offender and an identifiable victim of the resulting

[6]For a comprehensive discussion of the use of predictive sentencing, see De Keijser et al. (2019).

> crime, as well as an identifiable official at whom voters can direct their anger. The media and political opponents can ask, 'Why did you release Bob? Why did you expose Mary – here's a picture of her with her two cute children – to the risk of victimization?' (Pfaff, 2017, p. 168)

In other words, the false-positive problem consists in the fact that the emotional reactions to the two types of error are highly asymmetrical and that those who can be held responsible for the mistakes may well react accordingly. This is perhaps most significant in a system, such as the one in the US, where there are elected prosecutors and judges. As Pfaff puts it, judges and prosecutors may be "inclined to over-punish to avoid the risk of a false negative blowing up a re-election campaign" (Pfaff, 2017, p. 169). But politicians may also be affected along the same lines (Ryberg, 2020, p. 201).

The reasons that this problem is relevant in the present context is that it is not difficult to imagine precisely the same mechanism influencing decisions on crime preventive use of neurointerventions. First, with regard to the question as to whether a treatment should be initiated at all, there may be a strong inclination to seek to avoid false-negatives—cases where a treatment is not administered because an offender is incorrectly identified as not constituting a risk—but a comparatively weaker reason to try to avoid cases where the use of a treatment is inefficient because the offender would not have recidivated anyway. Second, once a treatment has been initiated, the same mechanism may also affect decisions on when it should be terminated. The risk of ending up with a false-positive outcome may provide a strong inclination to keep on treating an offender even when he or she no longer constitutes a risk. In public debate, it is certainly not difficult to imagine complaints along the following lines: "We know that this person has a criminal background and now he has committed a new crime. Why has this treatment been terminated when we know that it works? Who is responsible?" (see also Pfaff, 2017, p. 168; Ryberg, 2020, p. 202). Thus, in sum, in the same way as the false-positive problem may have contributed to an excessive use of punishment, it may also push decisions on the use of crime preventive neurointerventions in the direction of initiating treatments on more offenders than is necessary, and of abstaining from terminating treatment when it is no longer necessary. Both possibilities exemplify how a particular mechanism in real-life decision-making may lead to unjustified use of such interventions on offenders. As we shall now see, further reasons can be given to the same effect.

Ignorance of preferable alternatives. One of the criticisms that has repeatedly been advanced against mass incarceration in the US and elsewhere is not only—as we have seen—that offenders spend too long a time in prison, but also that it, at least for some offenders, would be preferable if they were punished in other ways. Not only have some criminologists underlined that the use of imprisonment may in some cases be counterproductive, in the sense that this punishment may increase rather than prevent involvement in future criminal activity, but even if there is a special crime preventive effect associated with the use of imprisonment, this type of punishment would still be morally wrong if there were other types of punishment that would result in a preferable balance between, on the one hand, crime prevention and, on the other, the adverse effects which the punishment has on the life of the offender. In short, penal practice suffers from the problem that one has often not chosen the best available methods to punitively deal with offenders. For instance, Michael Tonry summarizes his comparison between the current use of incarceration and possible alternatives along the following lines:

> There is good evidence that imprisonment fails to reduce later offending and may increase it. There is good evidence that well-designed, well-targeted, well-resourced, and well-run treatment programs can modestly reduce later offending. Treatment programs cost much less to operate than prisons and are less likely themselves seriously to damage offenders' likelihood of living better lives later on. (Tonry, 2016, p. 459)

The conclusion he draws is that many offenders should be diverted into alternative types of punishment, treatment programmes and community penalties. Noteworthily, though this conclusion may sound like one that would of course be endorsed if one held a consequentialist point of view, it is also one to which retributivists may subscribe. Even retributivists who believe that what matters morally is that offenders get the proportionate punishment could (and indeed should) hold that if two different types of punishment are equally severe—that is, if it is possible to maintain the "penal bite" of a punishment, then one should choose the type of punishment which has the best consequences (see, e.g., Ryberg, 2019). This is of course likely to be the one that will contribute the most to the prevention of future crimes. Thus, criticism concerning the current use of types of punishment need not presuppose a consequentialist approach to the justification of punishment. But why is this sort of criticism relevant in relation to the discussion of real-life use of neurointerventions?

The answer obviously is that if there is a tendency—whatever the underlying mechanisms may be—not to choose the *type* of reaction to crime that is ethically preferable, then this might also constitute a problem in the use of neurointerventions as a crime preventive instrument. Even if the administration of a particular type of neurointervention could in principle be morally justified, it would obviously be wrong to initiate such a treatment if there are alternatives that would have a greater crime preventive effect or have less deleterious implications for the life of the offender. For instance, this could be the case if programmes concerning the resocialization and reintegration of offenders into the community would have such comparatively preferable effects. Thus, the irrationality of current penal practice—i.e. the lack of use of the best available means—could also be feared to lead to an unacceptable use of neurointerventions (not least if, as has sometimes been underlined, it is the case that there is an immediate attraction in the idea of *curing* offenders of their propensity for crime).

Ignorance of academic information. A final, more overall, type of criticism that has often been presented against the US penal practice—and which in I my view may well also be raised against penal practice in many other countries—is that this practice is often the result of decisions which are not properly academically informed. For instance, commenting on some of the initiatives that have contributed to the excessive use of incarceration in the US, Tonry underlines this point in the following way:

> Mandatory minimum, three-strikes, truth-in-sentencing, 'dangerous offender', 'sexual psychopath' and LWOP[7] laws were enacted not on the basis of research findings, cost-benefit studies, impact projections, or meta-analyses, but because policy makers believed them to be intuitively plausible, morally appropriate, or politically expedient. (Tonry, 2016, p. 451)

But if mechanisms such as what is intuitively plausible, politically expedient and popular, means that criminal justice decisions are often not academically informed—in fact, there may sometimes (often?) be a direct political interest in disregarding academic information that is at odds with what is regarded as politically opportune—then it is not difficult to imagine that this tendency could also have serious consequences in relation to the use of neurointervention as a crime preventive tool. The proper administration of neurointerventions clearly requires that decision-making in various ways is responsive to scientific information. Adding such a treatment option into a system, which is driven by interests that do not foster such a responsiveness, may well have terrible consequences.

In summary, what I have suggested above is that, if one takes a close look at some of the trends and mechanisms that characterize current penal practice, then there might be reason to fear that these mechanisms would also influence the use of neurointerventions if such treatment was to be implemented within the framework of current penal practice. One way of underlining the discrepancy between ideal considerations and non-ideal theorizing is to note that the aspects of penal practice to which I have directed attention would usually not figure at all in ideal ethical considerations. In ethical discussions of neurointerventions, no one would serious raise the question as to whether such interventions should be used if they are pointless or if there are morally preferable alternatives. Neither would it be considered whether it is reasonable to administer such treatment without drawing on the requisite scientific knowledge. The answers to these questions may seem trivial. However, as indicated, once we leave the spheres of ideal theorizing and turn instead

7Life without parole.

to consideration of what we might expect if such treatment was to be implemented within the framework of current criminal justice practice, then the answers may well be far from trivial.

Current Use of Neurointerventions

The point in the previous section has been to consider some of the risks that may be involved in administering crime preventive neurointerventions within a criminal justice system such as the current one in the US which deviates significantly from what would be an ethically ideal system. Such considerations are of course important when one is assessing whether neurotechnological treatment that may become available in the not-too-distant future should be put into practice. However, there are also reasons to be cautious. Even if some of the above outlined mechanisms that are found in the way offenders are being punished exist and are even widespread, one cannot know for sure that they will also influence decision-making on the use of neurointerventions. In this sense, the considerations in the previous section might perhaps strike some as somewhat speculative. But are there any ways to examine whether these speculations can be buttressed?

One possibility might be to take a look at the way in which neurointerventions are already being administered in part of criminal justice practice. While most of the discussions of crime preventive use of neurointerventions involve hypothetical considerations of technologies that may be developed in future, there are some types of neurointervention that have already been put in practice in order to prevent offenders from falling back into crime. The most obvious example is the use of the chemical castration of sex offenders. This treatment works by reducing the testosterone level and thereby depriving an offender of most of the capacity to experience sexual desire and engage in sexual activity. There is variation between the drugs that are used in different countries. And the laws that have been passed authorizing the administration of this treatment also vary between jurisdictions. However, the overall question that could be raised is whether the concerns that I have raised in the previous section concerning the use of neurointerventions under non-ideal circumstances can be put to rest by examining the current use of chemical castration.[8]

The research on chemical castration supports the conclusion that such treatment does have a crime preventive effect. While studies have reached somewhat different results, the most comprehensive meta-meta-analyses on the effectiveness of such treatment have concluded that there is a positive effect with regard to the prevention of recidivism (Kim et al., 2016). However, obviously this does not answer the question as to whether such treatment is being administered in a morally acceptable way. Analogously, it may well be the case that punishment can serve both consequentialist and retributivist purposes, but clearly this does not show that punishments are being used in a way that is morally justified. As we have seen above, there are several reasons to believe that this is not the case. Thus, the more precise questions here is whether the patterns found in the use of criminal sanctions can also be found in the administration of chemical castration; namely, the tendency to overuse the treatment, to not administer the preferable treatment alternatives, and to rely on uninformed decision-making. A comprehensive review of the existing research would go far beyond the scope of this chapter. More modestly, I will direct attention to a few results indicating that it is not clear that the answer is in the negative.

First, when it comes to the overuse of chemical castration, this can happen in different ways. One way is to impose this treatment on offenders who do not constitute a risk; that is, who are not likely to re-offend. While it has been pointed out that some sexual offenders are indeed highly dangerous—such as those who are both paraphilic and anti-social—researchers have also underlined that some sex offenders do not suffer from sexual disorders but are nevertheless covered by castration statutes. At least some studies have indicated the risk of recidivism in this group of offenders to be low (see Kernsmith et al., 2016; Stinneford, 2006). Perhaps more importantly, it has also been underlined that there may be a tendency to over-treat some offenders. The risk that a sex

[8]For a more comprehensive discussion of the points in this section, see Ryberg (2020, pp. 210–213).

offender will recidivate may change over time. But some researchers have held that, once a treatment has been initiated, it is likely to be continued perhaps even for the rest of the life of the offender. For instance, Stinneford contends that "it is difficult to imagine a scenario in which the state, after deciding that an offender needed chemical castration, would later decide that it was 'no longer necessary'. Therefore, in most cases, the offender will be subjected to a life term of chemical castration" (2006, p. 580). To what extent such long-term treatment is justified or constitutes an example of over-treatment is an open question. But the question is of course relevant in the light of the side effects associated with this treatment.[9]

Second, though the question of the comparative assessment of different treatments in terms of effectiveness and side effects may not yet be sufficiently examined, there are some researchers who have commented on the relative merits of competing treatment option. For instance, Stinneford notes that, "several studies indicated that cognitive behavioural theory is as effective as chemical castration in preventing recidivism" (2006, p. 575). Though more recent studies have contested this view (e.g. Kim et al., 2016), it is noteworthy that these results have been reached several years after the treatment with chemical castration was introduced. This fact brings us to the final point.

Third, several researchers have underlined that decisions on the implementation of chemical castration treatment schemes have been made despite a widespread lack of sufficient knowledge with regard to several aspects of this treatment. For instance, in their recent review of the scientific literature on chemical castration, Rice and Harris have summarized their findings in the following way:

> Little is known about its effects on sexual or violent recidivism among sex offenders who do not freely request it. Moreover, little is known about the characteristics of those who volunteer for (rather than refuse) ADT [androgen deprivation therapy], especially such risk-relevant characteristics as psychopathy or high actuarial risk scores. Little is known about the

[9]Even if the long-term health effects of the use of chemical castration are not yet well-researched, there is a range of established harmful mental and physical health risks of chemical castration (see, e.g., Rice & Harris, 2011).

> long term effects of ADT on sexual behavior in general and sexual recidivism in particular. Little is known about the long term health effects of ADT. (Rice & Harris, 2011, p. 328)

It is remarkable that these general comments on the lacunas in the knowledge concerning the various effects of the use of chemical castration are presented more than a decade after the first US states authorized such treatment for sex offenders.

What these very brief considerations aspire to is nothing close to a genuine review of the existing research on chemical castration. Thus, strong conclusions are certainly not here warranted with regard to the assessment of the current use of chemical castration on sex offenders. However, much more modestly, the purpose has simply been to indicate that when it comes to the problems associated with decision-making on the punishment of offenders—namely, that there are tendencies to overuse this sanction, not to choose the preferable alternatives, and more generally to make uninformed decisions—and to the worry that these tendencies might also lead to unjustified use of neurointerventions, the current use of chemical castration cannot be held to completely dismantle these concerns. Furthermore, when it comes to the causes behind the unjustified use of the criminal sanction, namely, as several researchers have underlined, that criminal justice constitutes a highly politized field driven by various types of opportunistic political interests, the same point has been made in relation to the decisions concerning the implementation of chemical castration. For instance, commenting on the background of the current US castration laws, Rice and Harris contend that "the major impetus may have been political – sex offenders, especially paedophiles, are reviled by much of the lay public, and politicians who push for such laws clearly gain political favor by doing so" (Rice & Harris, 2011, p. 326). Thus, in sum, the current use of neurointerventions in the form of chemical castration does not provide sufficient ground for putting the previously outlined anxieties to rest.

The Significance of the Ideal/Non-ideal Distinction for Ethical Theorizing

Why is it important to keep the distinction between ideal and non-ideal theorizing in mind when it comes to the task of answering the initial question concerning the justified use of neurointerventions in crime prevention? There are, in my view, several answers to this question. The answer that has been given so far is that in approaching the question from a non-ideal perspective it is important to keep in focus the consequences that may follow if such treatment was to be implemented within the framework of an existing criminal justice system. However, there are several ways in which I believe it is important to keep the distinction in mind. As will be argued in the following, a failure to take this distinction into account may in various ways lead astray the ethical discussion of the use of neurointerventions. To see this, let us—in accordance with the previous considerations—refer to the discussion of whether it can in principle be acceptable (or required) to use neurointerventions in crime prevention as "ideal" considerations, and a discussion of whether it in practice—within the framework of the mechanisms that characterize and govern criminal justice—can be justified, as non-ideal considerations. As we will now see, there are at least four ways in which a failure to recognize the significance of these two levels of consideration may affect the quality of ethical discussion.

Sliding prematurely from ideal to non-ideal considerations. The first and most simple example of a failure to take the ideal/non-ideal distinction properly into account occurs when ideal considerations lead one to draw unjustified conclusions at the non-ideal level. Suppose, for instance, that it is the case that a particular type of neurointervention will have a crime preventive effect and that it, if properly administered, can be imposed on offenders without too severe side effects. In fact, it may even be imagined that the crime preventive effect could over time constitute a major benefit for the offender him/her-self because he or she will not run the risk of future imprisonment. Under such circumstances, it might seem tempting to conclude that the use of the neurointervention would be justified; that is, that its use in the real world would be morally desirable. However, the latter step is clearly premature.

Even if there are indeed circumstances under which the use of the neurointervention would—everything considered—be desirable, it might very well be the case that the conditions that would have to be met are not satisfied in reality. What we have seen in the previous section precisely is that, once we consider the use of neurointerventions within the framework of the actual societal and political context, there might be various mechanisms that would imply that neurointerventions might be used in unacceptable ways even if they could in principle be administered in a manner that is morally desirable. There could be mechanisms that would imply that such treatment would be overused that it would be administered in cases where one does not yet possess sufficient scientific knowledge, or where there exist alternatives to the treatment that are morally preferable. In my view, the history of the use of neurointerventions throughout the twentieth century constitutes a valuable source of information regarding the many things that can go wrong in the actual implementation and use of the treatment of offenders. There is little doubt that the use of neurointerventions, both in the form of psychopharmacology and psychosurgery, has a very sordid prehistory. Though it would clearly be premature to make inferences along the lines: "because a technology has previously been used in unacceptable ways, it will also be unacceptable to use it now", it is nevertheless the case that the prehistory of the use of neurointerventions can serve the function of opening our eyes to the fact that there is often a significant distance between ideal considerations and how a treatment technology ends up being used in a messy reality (see Ryberg, 2020, chap. 6). I tend to believe that a theoretical discussion of the use of neurointerventions does not always witness a sufficient awareness of how great the distance is between ideal considerations and real-life circumstances and, consequently, that there is a risk of sliding prematurely from premises of the former type to conclusions concerning the latter.

Discussing at different levels. Another simple way in which the failure to take the ideal/non-ideal distinction into account may lead ethical considerations astray is, of course, if two interlocutors fail to realize that they are discussing at different levels. This can happen in various ways. Here is an example which I believe can sometimes be found in ongoing discussion. Discussant A contends that neurointerventions should not be administered on offenders with a propensity for crime because this will involve a violation of the offenders' X—where X is some moral property that ought to be protected (e.g. autonomy, freedom, mental self-determination, dignity, etc.). Discussant B replies that the objection fails because the use of neurointerventions need not involve a violation of X. Either there are particular types of neurointervention that do not violate X or there are some neurointerventions which might potentially violate X, but which can nevertheless be administered in particular ways such that a violation will not occur. This exchange of arguments seems perfectly sound and, as noted, is familiar to anyone who takes part in academical neuroethical discussion. However, a failure to take the ideal-/non-ideal distinction may well imply that the exchange is confused.

If the discussion takes place at an ideal level, then A's objection might for instance be that in all cases in which a neurointervention is used it will involve a violation of X. And B can reply to this by showing that there is a possible world—which might well be hypothetical—in which the treatment would not violate X. Conversely, if the discussion takes place at a non-ideal level, then A's argument might be that, given the ways criminal justice practice currently functions and the way offenders are actually being treated, it is reasonable to believe that the implementation of neurointerventions as a crime preventive tool will involve a violation of X. To this, B might properly reply that given the way criminal justice practice works, there is still room for using neurointerventions in a way that does not violate X. These two types of academic exchange are of course totally sound. However, problems arise when the exchange involves a confusion of the two levels of discussion. If what A has in mind is the non-ideal argument, that given the way criminal justice practice works it is likely that the use of neurointerventions will lead to a violation of X, whereas B perceives this as an ideal argument and responds accordingly by depicting a hypothetical set of circumstances under which the use of neurointerventions would not violate X, then the discussion is clearly confused. B's contention simply does not constitute an objection to A's argument. It may be perfectly true that under actual conditions the use of neurointerventions would involve a violation of X, while also true that there are hypothetical circumstances under which this is not the case. While it might perhaps strike some as unlikely—at least when the exchange is spelled out as I have done here—that theorists would fail to recognize the different levels of the discussion, I tend to believe that there are discussions, where the ideal/non-ideal distinction is in this way not sufficiently carefully accounted for, and which consequently are led astray in the outlined manner.

Failure to consider the least evil. The previous examples of how a failure to fully take note of the significance of the ideal/non-ideal distinction may lead to premature conclusions or confused discussion both involving cases where the use of neurointerventions on offenders is justified in principle but may not be so in reality. However, the significance of the distinction is also witnessed in cases with the opposite starting point, namely, where the use of such an intervention is rejected as a matter of principle. Suppose again that we consider an argument to the effect that crime preventive use of a particular neurointervention is morally wrong, for instance, because this will interfere with a particular moral right (e.g. a right to mental self-determination). Suppose further that—unlike the situation in the above outlined argumentative exchange—it is the case that the neurointervention will always interfere with this right. There is no way to modify or supply the administration of the neurointervention in order to circumvent the right-violation. In that case, it seems *prima facie* legitimate to conclude that the use of such intervention would also be wrong in practice. However, on closer scrutiny, such an inference might nevertheless be premature. If, as I have suggested above, it is the case that the way offenders are currently being punitively treated in criminal justice practice deviates significantly from what is ideally desirable, then there is still room for the possibility that the use of a neurointervention would be morally preferable. It might be justified as the least evil.

Talking of the "least evil" may perhaps at first sight be associated with consequentialist reasoning. However, such a perspective might well also be relevant from a rights-based moral perspective. Suppose, for instance, that offenders are now being significantly over-punished but that it would in reality be the case, that they would be released much earlier if a treatment scheme involving neurointerventions was to be implemented. Now, even if, as assumed, it is the case that it has been shown that the neurointervention would constitute a right-interference it need not follow that the use of this treatment would be wrong under the outlined circumstances. For instance, it could be the case that a violation of retributive justice by punishing offenders in a disproportionate manner constitutes an even more serious moral problem than the interference with the right associated with the use of the neurointervention. Thus, if the neurointervention in practice constitutes the only alternative to over-punishment, this treatment may in this sense constitute the lesser evil and the option that should be pursued. It might also be the case that punishment beyond what is ethically justified is perceived as a violation of the offender's right—for instance, a right to freedom—and that this right is morally more weighty than the right that is infringed by administering the neurointervention on the offender. In that case, we would once again have a situation where the treatment by neurointervention might constitute the preferable alternative. Now, obviously the point here is not to defend a certain view of rights, but simply to stress the point that, if one as a theorist fails to recognize the significance of the ideal/non-ideal distinction by ignoring the fact that the actual ways offenders are being dealt with in criminal justice practice may differ significantly from what is morally desirable, then one might be misled to prematurely extrapolate a negative moral conclusion drawn at the ideal level to a conclusion on what should (not) be done in real life.

Missing the opportunity of real action-guidance. When philosophers and other theorists engage in ethical considerations, this activity is often justified by referring to the importance of guiding actions and decisions in a proper manner. However, quite often this justification does not mean the ethical considerations end up delivering very precise specification of which acts and decisions should be made. There are several reasons as to why this is so. For instance, it is often the case that philosophers outline the conditions that have to be satisfied for a practice to be justified, but where the question of when these conditions are satisfied in reality hinges on a number of empirical facts of which the philosopher is not well-informed. Or, if a certain practice should be governed by law, this precise question as to how these laws should be formulated are left to law scholars or other people with such an expertise. Thus, a well-known pattern is that philosophers provide the overall principles and that the final part of the work that is required in order to reach genuine action-guidance is left in the hands of other people who possess the requisite expertise with regard to empirical fact, formulation of laws or the like. Since action-guidance quite often presupposes expertise from various disciplines, there is nothing surprising in this type of division of labour. However, if there exists a major discrepancy between what is ideally desirable and the existing practice within a certain field, and if this discrepancy is not fully recognized by those engaged in the ethical considerations, then there is a risk that these considerations may ultimately be deprived of the possibility of delivering the most appropriate action-guidance.

Suppose, for instance, that criminal justice practice is dominated by decisions that deviate from what is ethically desirable by not being based on satisfactory ethical considerations on what justifies the punishment of offenders or because they are often academically informed. Suppose, further, as many social scientists and criminologists have suggested, that this is not basically due to a lack of insight by decision-makers but rather that the penal policy is a highly politicized field driven mainly by what decision-makers regard as politically opportune. Suppose, finally, that there are strong reasons to believe that some of the mechanisms dominating penal policy will also affect decisions on the use of neurointerventions on offenders. Under these conditions, the most proper way of ensuring that the treatment of offenders is carried out in a way that is as close as possible to what is ethically ideally desirable, need not be to convey the ethical conclusions in a direct way. Sometimes a conveyance of considerations that is adjusted to the obstacles that characterize a certain field may constitute a more effective way of diminishing the wrongs that are happening or of approximating the decisions that are right. This way of thinking draws on the well-known distinction in ethical theory between a criterion of rightness and a decision procedure. The distinction has traditionally been associated with consequentialist thinking. In a given context, the best way of ensuring the maximization of what is good might well be to follow non-consequentialist decision procedures. For instance, this would usually constitute the reason why consequentialists might subscribe to human rights or other like decision procedures that are functioning as better instrument to the maximization of the good than the preaching of the basic consequentialist criterion of rightness. Even though traditionally associated with consequentialist thinking, the same indirect way of thinking about action-guidance might be equally relevant for other ethical theories such as those that ultimately provide the most plausible answers concerning the crime preventive use of neurointerventions. However, in order to reconsider the ethical assessment of neurointerventions in a way that is designed to guide a reality that may not be dominated by open-minded and responsive decision-makers by various other political interests, it will be necessary to possess a close insight into the mechanisms that drives criminal justice decision-making. If the ethical considerations of philosophers and other theorists stay within the spheres of ideal theorizing and remain ignorant of the mechanisms that characterize the non-ideal reality, then the ethical considerations may well end up being deprived of real-life action-guidance (decision procedures) which, as noted, is usually emphasized as the justifying aim of engaging in such considerations in the first place.

Conclusion

The controversial question as to whether it can be morally justified to use neurointerventions on offenders as an instrument of crime prevention cannot, as initially pointed out, be answered without engaging in various conceptual clarifications of what precisely this question asks. What I have suggested in this chapter is that one important aspect that calls for clarification is whether the question should be interpreted as an "in principle" question, that is, as an invitation to considerations of whether it can ever be justified to administer this kind of treatment in crime prevention, or whether it should rather be understood as concerning whether it would be morally justified to implement such treatment under the circumstances that characterize current criminal justice practice. More precisely, it has been argued that some of the tendencies in the way offenders are being punitively treated by the criminal justice system—which differ significantly from what is ethically desirable—may also have implications for how offenders would be treated if neurointerventions were to be implemented as a crime preventive tool. Furthermore, it has briefly been indicated that the way in which some neurointerventions are currently being used—namely, in chemical castration—does not provide a firm ground for dismantling the worries concerning the use of neurointervention under non-ideal circumstances. Finally, it was suggested that awareness of the distinction between ideal and non-ideal considerations in a number of ways is important in order to avoid the ethical discussion of the use of neurointerventions being led astray.

An important implication of these considerations in my view is that they underline the significance of adopting an interdisciplinary approach to the initial question. Roughly speaking, philosophers tend to address the ethical question concerning the use of neurointervention as an ideal question, while social scientists and law scholars tend to interpret it as a question concerning the implementation under the actual circumstances. Both approaches are important. It is highly important to reflect on what basically matters and outline the ideal conditions that would have to be satisfied for the use of neurointerventions to be justified. However, it is equally important to be cognizant of the mechanisms that drive actual penal practice and to be aware of how this practice deviates significantly from what is ideally desirable. Thus, the overall conclusion is that the initially-posed question on the use of neurointerventions in crime prevention not only requires clarification along the lines of the distinction between ideal and non-ideal theorizing, but also underlines the significance of an interdisciplinary approach to the answer.

Bibliography

Fazel, S., et al. (2012). Use of risk assessment instruments to predict violence and antisocial behaviour in 73 samples involving 24827 people: Systematic review and meta-analysis. *British Medical Journal, 345,* 1–12.

Garland, D. (2001). Introduction: The meaning of mass imprisonment. *Punishment and Society, 3*(1), 5–7.

Husak, D. (2019). Why legal philosophers (including retributivists) should be less resistant to risk-based sentencing. In J. de Keijser, et al. (Eds.), *Predictive sentencing*. Oxford: Hart Publishing.

De Keijser, J., Roberts, J. V., & Ryberg, J. (2019). *Predictive sentencing*. Oxford: Hart Publishing.

Kernsmith, P., et al. (2016). Fear and misinformation as predictors of support for sex offender management policies. *Journal of Sociology and Social Welfare, 39,* 39–66.

Kim, B., et al. (2016). Sex offender recidivism revisited: Review of recent meta-analyses on the effects of sex offender treatment. *Trauma, Violence, and Abuse, 17,* 105–117.

Lacey, N. (2008). *The prisoners' dilemma.* Cambridge: Cambridge University Press.

Murphy, J. G. (1979). *Retribution, justice, and therapy.* Dordrecht: Reidel.

Pfaff, J. F. (2017). *Locked in.* New York: Basic Books.

Rice, M. E., & Harris, G. T. (2011). Is androgen deprivation therapy effective in the treatment of sex offenders? *Psychology, Public Policy, and Law, 17,* 315–332.

Ryberg, J. (2004). *The ethics of proportionate punishment: A critical investigation.* Dordrecht: Kluwer Academic Publishers.

Ryberg, J. (2017). Neuroscience, mind-reading, and mental privacy. *Res Publica, 23,* 197–211.

Ryberg, J. (2019). Risk and retribution: On the possibility of reconciling considerations of dangerousness and desert. In De Keijser, et al. (Eds.), *Predictive sentencing*. Oxford: Hart Publishing.

Ryberg, J. (2020). *Neurointerventions, crime, and punishment: Ethical considerations*. New York: Oxford University Press.

Singer, G. (1979). *Just deserts.* New York: Balling Publishing Company.

Stinneford, J. F. (2006). Incapacitation through maiming: Chemical castration, the eighth amendment, and the denial of human dignity. *University of St. Thomas Law Journal, 3,* 559–599.

Tonry, M. (2004). *Thinking about crime.* New York: Oxford University Press.

Tonry, M. (2009). Explanations of American punishment policies. *Punishment and Society, 11*(3), 377–394.

Tonry, M. (2016). Making American sentencing just, humane, and effective. *Crime and Justice, 46*(1), 441–504.

Valentini, L. (2012). Ideal vs. non-ideal theory: A conceptual map. *Philosophy Compass, 7,* 654–664.

von Hirsch, A. (1993). *Censure and sanctions.* Oxford: Clarendon Press.

von Hirsch, A., & Ashworth, A. (2005). *Proportionate sentencing*. Oxford: Oxford University Press.

Jesper Ryberg is Professor of Ethics and Philosophy of Law at the Department of Philosophy. He writes and teaches in the areas of ethics and philosophy of law. He is the head of the Research Group for Criminal Justice Ethics and is currently also head of the Neuroethics and Criminal Justice research project. Ryberg has published in philosophical journals such as The Philosophical Quarterly, Philosophical Papers, Theoria, Ethical Theory and Moral Practice, The Journal of Ethics, Res Publica, Journal of Medical Ethics, Neuroethics, Journal of Applied Philosophy, Social Theory and Practice, International Journal of Applied Philosophy, Criminal Law and Philosophy, Analysis, Utilitas, Ratio, and AJOB Neuroscience.

Neuroscience and the Moral Enhancement of Offenders: The Exceptionally 'Good' Brain as a Thought Experiment

Bebhinn Donnelly-Lazarov

Introduction

How blameworthy is a 'wrongdoer' with an associated brain 'abnormality'? Is it possible for the wrongdoer to be enhanced by biomedical interventions[1] and what would it mean, if anything, for that to happen? The fact that the blame-question can on occasion admit of opposite

I would like to thank Christopher Taggart, my colleague at the Surrey Centre for Law and Philosophy, for helping me to refine the chapter, and for a probing discussion of its premises.

[1] Douglas proposes a helpful, broadly applicable definition of moral enhancement: 'A person morally enhances herself if she alters herself in a way that may reasonably be expected to result in her having morally better future motives, taken in sum, than she would otherwise have had' (Douglas, 2008).

B. Donnelly-Lazarov (✉)
University of Surrey, Guildford, UK
e-mail: b.donnelly-lazarov@surrey.ac.uk

S. Ligthart et al. (eds.), *Neurolaw*, Palgrave Studies in Law, Neuroscience, and Human Behavior,
https://doi.org/10.1007/978-3-030-69277-3_10

answers (he is blameworthy because of his condition; he can be excused because of his condition), both plausible, indicates an intractability. This paper explores (a) the nature of the intractability, making the case that the issues are philosophical, not scientific, resolvable only by philosophy, and (b) the implications of intractability for the ethical contours of neuroenhancement. To assist the latter project, an explanatory shift is adopted. Rather than consider how a brain 'abnormality' affects or should affect our moral assessment of the wrongdoer, I ask how our response to the good person might change on learning that their unusual brain profile enhances their goodness. Is the person with an extremely 'good' brain morally better, or worse indeed, than the rest of us? What, if anything,[2] might it mean for us to be enhanced relative to the good person or for the 'good' person to be enhanced relative to us? What follows ethically for neuroenhancement and our approach to criminal offending? An insight, given added prominence by the change in focus, is that interventions have the potential not just to change the person, or to change the nature of human beings but to undo the existence of 'humans' entirely.[3] Axiological objections to neuroenhancement[4] miss the mark by understating the profound nature of change moral neuroenhancement may bring about, and in assuming epistemic access to a world so changed.

---

[2]The 'if anything' possibility should not be ignored. 'We can all agree that having certain altruistic or empathetic dispositions or less biased motives is a good thing. Nonetheless we can still contest the belief that enhancing those traits would make anyone more *moral*' (Melo-Martín, 2018).

[3]Some positions that may be thought similar can be distinguished. In particular the claim I make is compatible with the retention of key attributes often thought to be diminished by neurotechnological (particularly NCMBEs) or genetic interventions, including: agency (Kass, 2003; Sandel, 2007); the freedom to fall (Harris, 2011); autonomy; and authenticity (Sandel, 2007). For a nuanced discussion of various accounts of authenticity and the relationship to autonomy see Bublitz & Merkel 2009). Although the case cannot be made here, these attributes do not necessarily disappear even with the most intrusive of enhancements (though they may); the point rather is that *we* may disappear and epistemic access to the world of the new entity is lost. The risk may be thought remote but appear incrementally or by stealth. Note too that this risk appears whether enhancements are chosen or enforced.

[4]The term is used by Carter and Pritchard for a range of objections to their work (Carter & Pritchard, 2019). The thought behind axiological objections is that 'there is particular value associated with the kind of achievements that involve the overcoming of obstacles. Accordingly, by aiming to remove such obstacles entirely, some of the radical non-traditional forms of cognitive enhancement threaten to diminish a certain valuable dimension of human life'.

The Source of Intractability: Philosophy not Neuroscience

How ought we to view the 'wrongdoer' with an associated brain 'abnormality'? The matter is very much contested.[5] Consider the offender with a severe dangerous personality disorder. Some suggest that culpability is undermined by such a profile; that after all, this 'wrongdoer' is not truly bad. Others go so far as to say that their condition is the thing that founds culpability; that being like this is what it means (for this person) to be a bad person (Law Commission, 2013). The intractability is pervasive; we may equally claim, for the usually 'normal brain-profile' morally upstanding person, that an instance of criminal offending is out of character and so excuse it or insist that the agent is blameworthy precisely because they had the capacity to behave differently and chose not to.[6] We may even decide to excuse this offender, or not, because their brain state too explains their behaviour.

If there are difficult questions about the nature of an agent's moral position, naturally the associated questions about what it might mean to enhance that position are at least, and at least for this conceptually prior reason, equally difficult. The concerns of this paper, what has made these questions intractable, and the ensuing implications for neuroenhancement, emerge from well-worn, philosophical ground. And, some may assume that the matters are properly considered on philosophical ground. Others claim that these questions of old may be difficult for philosophers, but are increasingly clear to neuroscientists.[7] There is nothing persistent or intractable about them. Moreover, neuroscience may be

[5]This is true of legal systems. Consider Norwegian law at the time of the Breivik conviction where the insanity defence was based on a medical model, unpopular elsewhere. For a defence of medical models (see Moore, *The Quest for a Responsible Responsibility Test: Norwegian Insanity Law After Breivik*, 2015).

[6]Duff has long since urged a departure from the legal theoretical focus on choice or character as the locus of blame, convincingly deferring to actions themselves (Duff, 1993).

[7]For a comprehensive, critical account of various positions neuro-ethicists do or might take on these matters, see Racine et al. (2017).

self-sufficient in this respect; if philosophy finds the questions troubling, perhaps this is because they are just not its domain.[8]

What is the nature of the, apparently intractable, questions at issue? To attempt to understand blameworthiness and enhancement is to consider at least the following: what it means to be morally responsible; what makes a person culpable; which actions are culpable; what moral *improvement* might entail; whether such 'improvement' is *moral* improvement, whether it increases or undermines responsibility, and what kind of subject emerges from this process. Merely setting out the issues in this way makes the conclusion hard to resist that these questions are conceptual and moral ones, not scientific. Indeed, we may be caused to consider what it could mean to find answers to these questions in the brain. And, that consideration, too, is philosophical not scientific.

If this general position is correct, at what point does philosophy run out and neuroscience step in? Do the insights of neuroscience go at all to the intractable questions raised? Some are deeply sceptical about any such possibility (Berker, 2009).[9] The institution of law provides a useful tool to delineate matters neuroscience can shed light on from those it cannot. Consider the culpability question in criminal law: Is Bob culpable, partially culpable or not culpable at all for the theft of Bill's violin? In respect of the latter two possibilities, does Bob have an affirmative defence like duress, or a defence based on the absence of mens rea/actus reus? There is no doubt that neuroscience can assist to answer these kinds of questions. Brain-based lie detection, for example, may help asses a defendant's truthfulness and so test the reliability of the witness's

[8]Putnam is among those to say that this rejection of philosophy by neuroscience is at least sometimes an error: 'The idea that there is a scientific problem of "the nature of the mind" presupposes the picture of the mind and its thoughts, or "contents", as *objects* (so that investigating the nature of thought is just like investigating the nature of water or heat). I want to suggest that philosophy can expose this presupposition as a confusion' (Putnam, 2012). If mental contents are not legitimately pictured as objects, then the same must be true of moral being, comprised, at least in part, of mental content.

[9]The 'basic problem', according to Berker, is that 'once we rest our normative weight on an evaluation of the moral salience of the factors to which our deontological and consequentialist judgments are responding, we end up factoring out (no pun intended) any contribution that the psychological processes underlying those judgments might make to our evaluation of the judgments in question' (Berker, 2009, 327).

evidence, or so it is claimed.[10] PET or fMRI scans may tell us something about Bob's likely capacity to plan or form intentions and so undermine (or advance) the prosecution case.[11] If the neuroscientist can show that Bob does lack these capacities, from a legal perspective the conclusion may follow that there was no mens rea and so no guilt. Or, the science might tend to support a defence position that even though the defendant ostensibly committed the crime, he was having a seizure brought about by concussion at the time and so, at the relevant moment, 'he' was not truly present at all. A defence of insane automatism is in this way supported.[12] It is vital to understand in the court room, and for a variety of legal reasons, whether the defendant really was concussed, what affect such concussion could have on their behaviour, that he had a brain tumour affecting libido control at the time of the 'offence', that she was heavily intoxicated, or suffers from a personality disorder. More radically, neuroscientists may claim to be able to show what mental state the defendant was in at the time the offence was committed or to distinguish among mental states. Since this is likely to be a key matter for the most difficult criminal appeals—whether the defendant intended, knew, was reckless, foresaw, suspected or was entirely ignorant of the likely consequences and or circumstances of his actions—the potential utility for law here is great.[13]

---

[10]The limitations of brain-based lie detection are considerable and not merely scientific. Patterson and Pardo note 'It is a conceptual mistake to assume that brain-based lie detection provides direct access to lies, deception or knowledge'. More fundamentally, the authors propose that 'People, not their brains, lie, deceive, know and remember' (Patterson & Pardo, 2013, pp. 105–106).

[11]The M'Naughton test for insanity will be satisfied where: '... at the time of the committing of the act, the party accused was labouring under such a defect of reason, from disease of the mind, as not to know the nature and quality of the act he was doing; or, if he did know it, that he did not know he was doing what was wrong' (R v M'Naghten [1843] 8 E.R. 718). The test has admitted of various interpretations across time and jurisdictions. The fact that this is so indicates not that neuroscience can provide certainty but that the neuroscientist cannot be clear what she is expected to look for in the first place.

[12]Providing problems of contemporaneity can be overcome.

[13]More generally, as Racine et al. claim, 'it is hard to deny that ethical theories would benefit from an up-to-date understanding of the biological and psychological underpinnings of moral judgment' (Racine et al., 2017). In noting the limitations of mental-state analysis, again Patterson and Pardo make a useful contribution: "One makes no sense in saying; 'I intend X and X is impossible' or in claiming, 'I know Y and Y is false.' Yet, if intending and knowing were simply brain states, the statements can stand. It follows that neither can be a brain state"

Neuroscience may also help when it comes to decisions about how to treat or punish offenders, complementing and apparently adding precision to the 'softer' conclusions of probation reports, pleas in mitigation, and the judge's experienced conclusions. It may claim to shed light on the nature of decision-making about punishment, exposing bias, identifying the emotional components of these decisions and the like, though some are sceptical (Patterson & Pardo, 2013, pp. 186–191). And, if we want to deter future offending, to suppress it, or to encourage a better ability to reason with and through moral norms, neuroscience can suggest tailored cognitive and non-cognitive interventions: antidepressants, oxytocin, SSRIs, Depo-Provera (reducing aggression, impulsive behaviour, sexual offending, poor empathy), genetic interventions, and brain stimulation. Notwithstanding the complex interaction of ethical, empirical, and scientific barriers that attend the same, these interventions at least suggest alternatives to society's atavistic, and for most purposes ineffective, reliance on imprisonment.

So there is no doubt that neuroscience can assist law and help to answer some of its important questions. What neuroscience cannot do is determine the questions that warrant asking. Its role is never a foundational normative one but a secondary practical one. In this way, neuroscience does not tell us whether the posited law of theft is as it should be,[14] what it means for Bob to be responsible, whether Bob ought to be punished; what it means conceptually to lie, how intention should be defined in law, whether biomedical interventions are justified. It is once we switch to the normative and conceptual questions that science faces limits. To understand what it means to kill intentionally is not, ever, to look in the brain of an offender.[15] Nor, if we want to find out whether

(Patterson & Pardo, 2013). A very recent contribution to the debate purports to distinguish knowing from reckless mental states (Jones et al., 2020).

[14]This is not a matter that science can assist with. Moore makes the point: 'The law must define legal concepts for itself in light of legal purposes. The law cannot simply adopt a concept developed by psychiatrists for therapeutic purposes, or for that matter any concept developed by any social scientists for explanatory purposes. The purposes of the law in question must govern the definition of any term appearing in that law; no other discipline's conceptualization can safely be adopted and plugged into a legal formula' (Moore, 1979).

[15]This is a particularly important limitation because in recent times the typical formulation of action as 'willed bodily movement' has faced sustained criticism. If neuroscience is

the person with a personality disorder should be excused for their 'wrongdoing', will the brain enlighten.[16] Neuroscience cannot itself speak to: what human action is, which actions ought to be criminalized, which mental states ought to be exculpatory or inculpatory, how to punish if at all, how law ought normatively to rank the culpability of particular mental states, what it means to be morally enhanced, and whether that enhancement is justified. These limitations remain present however sophisticated our knowledge of brain mechanisms might become. So, if we were to write criminal law anew and to redesign systems of punishment in light of the findings of neuroscience, and in order for law better to reflect moral responsibility, those findings can only be of secondary importance. This is not to say that this secondary importance is insignificant. It is very significant indeed but a proper delineation of explanatory roles is required for the contribution of neuroscience to be helpful. And once more, it is again an ethical, not scientific question whether and how neuroscience ought to proceed with interventions in the face of intractable problems it cannot itself solve.

An Alternative View

Some of the proponents of neuroenhancement are occasionally drawn to an alternative account of these issues such that even the manner of presentation set out above is disputed. In their story, one 'old' responsibility question is answered straightforwardly and in deterministic, biological terms. Agents are not responsible, brains are. There is no

---

to assist us to understand whether a brain state correlates with or amounts to an intention, clarity is needed about what it means to intend, to act intentionally and to act with an intention. Anscombian accounts pose particular difficulties for any neuroscience of intention (See, for example, Donnelly-Lazarov, 'Intention in Criminal Law: The Challenge from Non-Observational Knowledge', 30 Ratio Juris 4, 2017).

[16]Moore puts the point well in relation to the function of neuro medical analysis: 'Given the explanatory, curative, and preventative purposes behind medicine's taxonomy of diseases, there is little reason to expect responsibility to turn on whether an accused has one disease rather than another. Granted, to explain the accused's condition, and thereby to be in a position to cure it or to prevent its recurrence, such knowledge is indispensable to medical practitioners. But such particular understanding (in terms of the medical nosology) is by the-by when it comes to evaluating whether an accused is responsible or excused for his behaviour' (Moore, *The Quest for a Responsible Responsibility Test: Norwegian Insanity Law After Breivik*, 2015).

place for punishment or condemnation; the role of neuroenhancement is simply to respond to offending by occasioning utilitarian benefits.[17] This position is a radical one, consigning difficult responsibility questions to non-existence; we need only be concerned with cause and effect. And since we need only be so concerned the position is considered neutral; it does not proceed from moral judgement, no such thing is warranted, but from apparent facts about the brain.

In truth, the lack of neutrality here is easily exposed. If neuroenhancement is legitimate in virtue of its capacity to affect a change in patterns of offending and if this is a scientific truth, in the sense that it proceeds from matters of fact, it should be immune to theoretical challenge. But the theoretical challenge arises just because we all do and must consider how to behave.[18] Self-reflection—including reflection on our moral capacities—just is a feature of our human nature. We are beings, possessed of reason, who must act in a world inhabited by other such beings. Therefore, asking, responding, and acting through ought-questions is a feature of the human condition as much as hunger or sight or movement, or the possession of a brain.[19] Even in the unlikely event that all these normative engagements can be reduced *to* the brain, still, it is an undoable fact that the processes present themselves to us. Why does this bear upon the position of the neuroscientist who wants to say that brains are responsible, not us? It affects that position because if brains are responsible, not us, normativity is flattened; only the 'is' survives. There is no way in which we ought to have acted for we could only have acted the way we did. Hard determinists happily agree. But, notice that this

[17]Racine et al., note the trends (Racine et al., 2017). Recently, Caruso has argued for the role of neurosciences in bolstering determinism (Caruso, 2020). See also (Cushman et al., 2010; Singer, 2005) Patterson and Pardo provide a counterpoint: "[Greene] attacks deontology (and, by a loose extension, retributivism) for not having access to some 'independent [moral] truth', but this is precisely the kind of access he would need to impugn the decisions implied by a retributivist theory" (Patterson & Pardo, 2013, p. 190).

[18]Even when we do not reflect consciously on our actions—habitually making coffee each morning—we are guided by norms.

[19]Aristotelian and constituitive theories are particularly strong in pressing the point: 'human beings are condemned to choice and action. Maybe you think you can avoid it, by resolutely standing still, refusing to act, refusing to move. But it's no use, for that will be something you have chosen to do, and then you will have acted after all. Choosing not to act makes not acting a kind of action, makes it something that you do' (Korsgaard, 2008, p. 8).

is not a conclusion merely about practices of blaming or about culpability. It extends to oughtness itself. One cannot consistently claim both (a) that Bob cannot be condemned for stealing Bill's violin because he could not have done otherwise and (b) nonetheless we are compelled to ask normative questions about Bob's moral status and how to respond to it. The inconsistency lies in the fact that if brains determine what we do then we will simply respond to Bill the way we do respond; there is no oughtness to the matter, no need to consider whether to be a utilitarian or not. If there is no *real* choice for Bob whether to take, or not to take, the sought-after violin, there is equally no real choice for us about 'how we ought to respond' in our broad moral enterprise. Normativity is destroyed all the way down. (And so any theorizing about that normativity must be a false-enterprise.) The untenable implication of this position is that there is no need to be the kind of beings we are; to ask the kind of questions we do. And now too the proper subject of neuroscience (us) disappears. The 'we' investigated is no longer the normative human being, nor, so, is the brain of that imagined entity the brain 'we' have.

The conceptual and ethical contours of neuroenhancement need to be considered philosophically in order that the science is appropriately guided. What kind of guide can philosophy provide? The typical approach to these kinds of questions is to consider the bad person; the psychopath, say, who may be subjected to a biomedical intervention, inhibiting his propensity to offend or enhancing his capacity for moral engagement. Even if we can straightforwardly claim that such an intervention will bring about certain utilitarian benefits, there are vast areas of intractability: what is the agent's pre-enhancement moral position (was he truly blameworthy or not in the first place); what are the moral and existential effects if any of intervening; is enhancement justified. A change in focus might make some areas less grey.

The Good Person

It is not surprising that philosophical responses to culpability and enhancement questions turn to the behaviour of the 'wrongdoer'.[20] First, the questions are often asked from the perspective of legal philosophy, where the subject is an offender, someone who has transgressed, and institutionally it is in the criminal justice system that enhancement is likely to take hold (Shaw, 2017).[21] Second, 'wrongdoing' naturally is the form of action that causes the greatest moral concern, and we may fairly expect a person's wrongdoing to reflect their moral position. (If Bob steals his neighbour's car, we can confidently say that he is blameworthy; he can properly be condemned for what he did.) Third, a focus on wrongdoing brings to light those occasions when this expected relationship between behaviour and moral position breaks down. Particular brain states might make it so, and we may accordingly (or not) take a different moral view of the thief who has a long-standing mental condition, a temporary such condition, who is concussed, under extreme pressure, intoxicated, or whose behaviour is a rare departure from a typically moral life. Finally, if any such clear or even blurred lines can be drawn, it is surely the wrongdoers we might want to enhance not the good-doers.

But, notwithstanding the clarity of focus and sensible explanatory strategy at work, the 'bad man' has not yielded clarity of insight. A change in focus from the wrongdoer to the good-doer might expose something new or help disclose why the questions posed have proved intractable. What view do we take (and should we), of the good person whose goodness is a function of a 'good brain'? Should we (or they) be enhanced depending on the conclusions we draw and what might it mean, if anything, for this to happen? Some indication of the possibility for new insight comes from our intuitive response to these questions; if we are told about their brain condition, and actually asked the question; 'so, what do you think of this person now?!' we may very well (feeling empathy with the good person) find the question plain rude. We might

[20]Typically the psychopath. See, recently Baccarini and Malatesi (2017).

[21]Shaw imagines an offender "provided with a 'reform pill', which significantly weakened his desire to reoffend" and asks "Is the offender's subsequent, apparently 'good' behaviour genuinely good?" (Shaw, 2017).

at least find it odd, tending to think that human beings are often the people we take them to be; minimally that 'taking them to be' is the only way to know a person at all: we form views about people based on what we see, hear and feel, being so exposed to the sincerity of their efforts; the degree of self-sacrifice they make; the extent to which they consider the implications of their actions on the lives of others; and to their more instinctive, unreflective responses to moral events.

If the spectre of neuroenhancement was then suggested as a possibility for us, we may be even more horrified,[22] fearing an existential challenge to our nature, however imperfect we believe ourselves to be. Our intuitions may be wrong of course, but the very fact that the questions unsettle, might cause us to wonder why the parallel wrongdoer questions have at least become less unsettling. So, an examination of the good person exposes certain base intuitions. Do these, then, have any claim to theoretical soundness?

Imagine a person who inspires in us nothing but admiration. Who is generous, caring, careful, sincere, selfless. Imagine that they have these traits to a degree that must surely be exceptional. When we interact with them, we cannot but feel admiration for their great character and perhaps some shame for our own. If we were caused to assess the person in terms of their moral compass, we would rank them among the very, very best. We might be envious, look for flaws, be sceptical that anyone could be so wholesome but, in the end, we know, we really know, that it is ourselves who just can't measure up. This is a person who warrants the admiration they receive.

Consider the actions such a person performs: assisting those in need having carefully reflected on the morally most supportable way to do so; taking a brave and difficult stand against a popular but dangerous orthodoxy; coming to the assistance of drowning strangers while risking their own safety. And, jumping in front of the runaway trolley so that it kills no one. We would likely have no difficulty in concluding that this superhuman can be praised for these great acts. That they are responsible for what they do, and the doings are very creditworthy indeed. But, something happens that is to shake our certitude. Our hero has

---

[22]Some empirical support can be gleaned from (Specker et al., 2017).

an accident, not something dreadful but sufficiently serious to warrant a precautionary brain scan. Thankfully the accident has no serious effects. But, on examination, it turns out that all the areas of the brain associated with moral cognition, appetite, practical reasoning, empathy, and commitment are unusual. Our superhero has what we might call a super-normal brain. In particular he has enhanced capacity to reason and act well. Should we come to know that this is the case, how ought we to respond?

There are at least three possible theoretical responses to the new information: we learn nothing new about the person; something new about the person; or something new just about their brain:

(1) We might think we learn nothing new about the person, if we acknowledge; okay, this person is good, at least partly, because their brain is good. I am morally mediocre, or flawed, or worse, most likely because my brain is that way too.
(2) We might think we do learn something new if our response is to say, 'this person is not good at all. It is their brain that is doing all the moral work – not them. I have to work at being good, and boy is it hard!'
(3) We might think we learn something new only about the brain if our response is simply, 'this is what this part of her brain looks like'. But, so what, goodness is not about the brain nor does it reside therein?[23]

Each possibility has radically different implications for how we understand the moral position of the good-brained person. We either think that their good nature is accounted for by their brain state, that their ostensible goodness is exposed as a sham, or that the brain scan speaks to the nature of the agent's brain and not at all to their moral character. It might also be considered that each has entirely distinct implications for neuroenhancement, its justifiability, and its nature. In fact here things are not so straightforward. Manifold and shared implications are suggested

[23]This might be understood as an iteration of the mereological fallacy, a concept at the core of Patterson and Pardo's work (Patterson & Pardo, 2013).

across the categories. Consider the first; in respect of it we might say any of the following;

- Since we now know what a good brain looks like, let us all be good. To optimize our shared moral experience is something that humanity must strive for. So, we should endeavour to change our brains and develop interventions that can enhance our individual and collective goodness. Indeed, such is the dire state of humanity that we must make this an imperative.[24]
- But, for a number of reasons, we might equally respond, let's do nothing. Perhaps we think there is virtue in a society where flawed human beings act in the difficult pursuit of human goods; or that there is a good in imperfection; and that we should have the freedom to fall.[25]
- We may go much further still. Perhaps the processes of moral learning that we go through when we observe the actions of admirable people (whatever the source of their admirable qualities) and our striving for a more fulfilling way of being are not just beneficial; rather they are entailed in what it means to be the kind of, human, beings we are. Interfering with these sorts of normative relations is interfering with the 'oughtness' that characterizes human nature.[26] Enhancement portends not a change to human nature but an abolition of the human. The fear may be mistaken but if it is not 'us' that is to survive such enhancement, how would 'we' go about finding out? The implications of interference are an ontological and epistemic minefield; the attended existential risk very high indeed.[27]

---

[24]A point notably pressed by Savulescu and Persson. See for example Savulescu and Persson (2012).

[25]For variations on these ideas, see Jotterand (2014), Harris (2011), Parens (2005), and Sparrow (2014).

[26]Buchanan is among those to be sceptical about such appeals (Buchanan, 2011). He understands human beings to 'possess a conception of the good by which we can and do evaluate human nature' (115). This misses the incrementally arising risk that 'we' would no longer be doing the valuing. The new 'we' may not have any such conception.

[27]Fukuyama's concerns about a post-human world assume our ability to engage with it (Fukuyama, 2004). In truth it may not be possible to sketch any picture of that world.

Some of the options for the theorist who thinks he has learned something new about his friend are entirely compatible with these.

- For example, the conclusion may appeal that even though there is no true virtue in their 'goodness', the actions of the good person do no harm and therefore there is no justification for neuroenhancing the 'good man'. In fact, under a utilitarian calculus, very likely these actions have a net worth. It would be perverse to intervene to reduce these benefits.
- Or one may reach the same conclusion from the existential threat posed above; by accepting that we and our societies are composed of those who do good with great moral striving and those who cannot help but do so. To alter this is to alter something quite profound.
- Alternatively, from the view that the super good person cannot avoid being 'good' whereas the normal, regular good person has a lot of moral work to do, we might conclude that after all the latter is the good person. That the 'normal' good person deserves moral credit when they live up to our expectations. If we have a rewards system for creditworthy actions, much like our system of punishment for offending, we might wish to disqualify from it those with super-normal brains.[28] We might do more than this. We might say, 'OK, we will give you a chance but in order to see if you are really good, take this drug, one that over time will produce a normalizing effect in your brain. Then come back, let us see how good you really are?'

Only the 'so what' category seems to admit of a single response in terms of what follows for neuroenhancement:

- Part of the brain of the good person is unusual. The same part of the 'normal' person's brain is not. But, goodness is not about brains; it cannot be reduced to specific neural characteristics. People act, not parts of brains. Sure 'enhancing' the person will do something to the subject but it will not affect her *moral* being.

---

[28] Financial rewards in the workplace, honours systems, grants to support charitable causes might be such things.

Slightly Less Intractable for the 'Good Person'

The analysis above might seem to suggest greater, not less, intractability. At the same time, it has brought some of the more fundamental, risk-attending potentials of moral enhancement into view; an existential threat in particular. That threat, in turn, is in keeping with our intuitive response that there is something deeply worrying about neuroenhancement. In essence, the 'good man' analysis tells us something about blame and our responses to it that 'bad man' analysis tends to miss.

Why? A focus on the bad man proceeds from the assumption that it is necessary to consider whether brain states can be exculpatory and to assess whether some states differ from others in this regard. This is the explanatory design. But the design is mistaken. It is not possible to say whether brain states are inculpatory or exculpatory because our brain will always be in some state, a physical one, when we act. Nothing normative is suggested. This accounts for intractability; for our not being able, from brain state-based premises, to defend the view say that psychopathy inculpates, or exculpates. It may seem that this conclusion is too strong. Surely, for example, neuroscience can expose a lack of capacity and in this way offer a basis for exculpatory claims. But even in this regard the conclusion needs first to be defended, however obvious it appears, that a lack of capacity does indeed exculpate. The only way in which we can address inculpation and exculpation is by moral analysis of the actions people do. Why do we give a defence to the concussed? Not because his brain is in a particular state but because 'he' is absent. And we need to make the case that this absence matters for culpability. Neuroscience, of course, might confirm the evidential likelihood of the absence but does not speak to its normative relevance. Why do we not usually excuse the intoxicated? Not because his brain state is normal (it isn't) but because 'he' at an earlier point took a decision to reduce his capacities for foresight, and self-control and so can be blamed accordingly. Why are we reluctant to excuse the person with a dangerous personality disorder? Because we tend to believe that 'he' remains morally present in the actions done, actions we are sure ought not to be done. Why might we be sympathetic to the homeless offender who has committed a minor offence? Because of the conditions of his life and how we reckon we might be affected in similar circumstances. Why might we even feel some, if not great, sympathy for the hot-headed, reckless young man who punches another in a moment of angst? Not because his brain predisposes him to behave in this way (brain states always pre-dispose) but because we may feel he is lacking in maturity, but lacking also in hatred, likely to respond well to good counsel, and that these things matter, at least somewhat, in our practices of blaming and excusing.[29]

We come to know all of this because of what we understand about actions and actors, as one of the same. And these are matters that can only be observed and understood in the world, there being no place in the brain that moral action or moral interaction resides.[30] To know something about the brain, is to understand vital correlations, but it is not to know more about moral character and behaviour than what observation tells us, for it is not to know these things at all. Brain states just are. Moore captures the intuitive thought 'no one is morally blameworthy for his disabilities',[31] and Moore would no doubt agree that, 'no one is morally blameworthy for the state of their brain whatever that may

[29]Again Moore's contribution to a related debate is insightful: 'Thus when a person with a cold shoplifts the medicine he needs to cure it, he is indicted for the act of theft, not for the having of a cold. That the cold is not his fault is thus by-the-by—unless we can show that the having of a cold excuses his act'. Moore also notes that if we attribute relevance to brain states as such we need to do so universally and be careful in the kind of relevance we attach. Causation, for example, does not get us there: 'Causation of behaviour cannot excuse the mentally ill without also excusing all of us, healthy and ill alike. If it is the case that a disease prevents an actor from doing other than he did, then it needs to be shown why a non-diseased brain does not also prevent an actor from doing other than he did' (see Moore, *The Quest for a Responsible Responsibility Test: Norwegian Insanity Law After Breivik*, 2015, p. 666). This is not to say that brain states are normatively irrelevant. It is to say that we always have some normative work to do in the world.

[30]Neither does each combined and entirely unique exercise of mental and physical capacities that constitute moral action, nor the process of these causing us to recognise our actions as such, exist therein. The highly individual nature of action makes Douglas's hypothetical too much of an explanatory convenience that: 'The *only* effects of Smith's intervention will be (a) to alter Smith's psychology in those (and only those) ways necessary to bring it about that he expectably has better post-T motives, and (b) consequences of these psychological changes'. Since we never operate with discrete mental states there is no way in which a change in our psychology could be so individuated (Douglas, 2008).

[31]Moore, *The Quest for a Responsible Responsibility Test: Norwegian Insanity Law After Breivik* (2015, p. 664).

be'.[32] If we are asking whether brains states are inculpatory or exculpatory, we are asking questions with no answers. If we are asking whether it is right to alter the brain to make the person better, we misunderstand what being better means.[33] If, on the other hand, we remain attentive to actors and what they do, to their and our messy intricate realities, making assessments refined by experience, reflection, empathy, and compatible with the systems of reasoning philosophy provides, we may be more cautious about enhancement projects. We may so be more inclined to predict the damage occasioned to an infinitely complex moral world by even simple tinkering.

Implications for Enhancement of Offenders

We are somewhat epistemically care free when it comes to goodness. If a stranger's actions bear positively upon us, we are unlikely greatly to scrutinize their character. If someone we know impresses us morally, we likely accept that they are good. And, if we are unsure, it does not occur to us to find out about a person's brain to resolve any concerns, any more than we would want to view our own brain to answer a question put about an aspect of our own character. We are less care free when it comes to the bad man. This may follow for a number of reasons; first, in relation to the culpability question, we do not want to condemn a person without being *sure* we understand his action, his reasons for it, and the factors that influenced his decision-making. This understandable caution may cause us to challenge our natural forms of assessment and certainly to see the appeal of science. We may be inclined to think that, whatever our intuition, a person should be excused if his brain was 'faulty', that if their brain can be enhanced for the benefit of society, the opportunity wholeheartedly should be embraced. The earlier analysis indicates

[32]Even in the case of an intoxicated offender, we might say that he can be blamed for what he does when intoxicated but not for the state his brain happens to be in at the time—the state of his brain is simply a fact, in itself of no normative import.

[33]Of course, this is not to say that enhancements have no effect. Still, it is the *person* not a particular part of their brain who must act and we know their moral status based on how they do act.

that the first conclusion lacks the empathy it might appear to have for it replaces the person with the brain. (To observe the brain is not to understand the person or their actions.) The second lacks the good sense it projects for it proceeds from this mistake and ignores existential threats entirely.

When we consider the ethics of moral enhancement in the context of offenders, utilitarianism has a natural appeal. Even amid pressing and complex ethical concerns, the potential to bring about a dramatic reduction in the harm humans experience persuades almost in itself. In the absence of the reduction-in-harm imperative, (a perspective that the good brain' allows) the moral, conceptual, and ontological spectre of moral enhancement is, on more neutral ground, exposed. To understand the ethical contours of moral enhancement is to have sight of all the possibilities that attend it. The focus on the actions of the bad man gives artificial prominence to some; the potential to reduce harm, the potential to bring about genuine moral improvement, and the potential, for manifold reasons, to fail in the same. The intuition that interference is more profoundly troubling reflects the alternative possibility that moral enhancement radically might change what it is to be a person.

Imagine a world leader who cares not a jot about climate change, who wages wars, restrains our liberties, condemns minor offenders to death, develops more efficient WMDs, who himself flagrantly offends against the criminal law, who risks our well-being in weird and ever more destructive ways. Imagine now that the leader is morally enhanced, along with his coterie, and ministers, and their departments, and the electorate too—so that they no longer elect such leaders (who in any case would no longer be).[34] What a wonderful world? What would such a world be like? Well, we will have absolutely no epistemic access to that world for entities like us will not be in it.[35] And since the question put admits of no answer, it is a world full of risk. A place of beings

[34]I have in mind here the kind of social-policy-based enhancements that are the subject of some of Buchanan's thought (Buchanan, 2011).

[35]This epistemic closure makes it clear that what is at stake is not 'whether, or how much, normative weight to assign features of the human condition that have traditionally been taken as given….' (Juengst, 2019), it is the human condition itself. The conclusion also challenges Douglas's confidence that, 'On any plausible moral theory, a person's having morally better motives will tend to be to the advantage of others' (Douglas, 2008, p. 230). No doubt this is

with enhanced empathy, trust, self-control, powers of practical reasoning, ability to universalize, honesty, self and mutual respect, with reasoning processes playing an entirely transformed role in 'moral' life, would be nothing like the world we have.[36] Such beings would be nothing like us. We cannot say what these levelled[37] creatures would behave like, what priorities they would have, and what new moral landscape and moral challenges would emerge.

Conclusion

The field of moral enhancement is often the field of deflation. We are urged to keep the waters cool, advised that moral enhancement is comparable to long-standing, uncontroversial educative steps, certainly not the stuff of high tragedy some present it to be. But, the field is equally characterized by grandiose claims. We are to proceed with enhancement or the future of humanity is at stake![38] The spectre of the good brain helps expose the folly in both positions. Of course some enhancements have modest and useful impacts,[39] but others have the propensity not merely

---

very often true but it is not necessarily true of all, and in particular of extensively implemented, enhancements.

[36]It might be said that environmental interventions might occasion the same 'harms'. In the view of this author, some environmental interventions are worrying in some of the same ways as neuroenhancements but not in this particular way.

[37]Levelling may occur because beings like this probably have no need to exercise those capacities that enable and inhibit moral being. But, then again, who can say?

[38]For example Persson and Savulescu (2008). The authors say, 'What constitutes moral enhancement will depend on the account one accepts of right action'. Offering none they urge us to proceed as a matter of urgency, rejecting the modest 'freedom to fall' objection as hyperbolic. There is surely irony in their claim that, 'The expansion of our powers of action as the result of technological progress must be balanced by a moral enhancement on our part'. The most damaging expansion in our powers of action would be the undertaking and undergoing of mass moral enhancement. Melo-Martin makes a similar point: 'This makes all the more puzzling the insistence that yet more technology can save us from our folly' (Melo-Martín, 2018).

[39]Indeed, the fact that we cannot predict how much tinkering, of what sort, when, or how it might lead us no longer to be us, is part of the problem. See Pugh for a clear articulation of the parity objection. (The objection is undermined if the incremental story mooted here has *moral* import. Indeed, that incremental story does not commit to the view that all environmental interventions are permissible.): 'Establishing that NCMBEs violate freedom may in fact be counter-productive to supporting one's opposition to NCMBEs, if one does not supplement this

to affect moral outcomes or capacities but also to something unknown and unknowable. [40] So, it is vital to consider not only whether the 'enhanced' person with poor self-control is thereafter morally the same, morally improved, or just the opposite, but also whether enhancement amounts to something else entirely; an existential threat. It is equally important to consider this in collective as well as individual contexts. A minor enhancement may be insignificant for an individual in isolation, but very significant indeed if undertaken en masse. If, for example, 'the people transferred their shallow values to their children, humanity could get permanently stuck in a not-very-good state' (Bostrom, 2003). The point pressed here is that depending on the nature and direction of any such process, it may not be humanity, but something else entirely, that is so 'stuck'.

The grand claim too fails. We should not proceed apace with enhancement precisely because if we do the future of humanity is at stake. Undoing those things that bring misery to human beings is not freeing human beings from misery, it is creating another sort of being. To be concerned for the future of humanity is to be concerned as the beings we are for the beings we are. It may well be good to proceed to this world and that is a case worth exploring. What cannot be said with any certainty is that it is good for 'us'.

Bibliography

Baccarini, E., & Malatesi, L. (2017). The moral bioenhancement of psychopaths. *Journal of Medical Ethics*, 697.

Berker, S. (2009). The normative insignificance of neuroscience. *Philosophy and Public Affairs*, 293.

argument with a plausible account of why seemingly plausible environmental interventions that directly modulate emotions or behaviour differ from Moral Bio-enhancement' (Pugh, 2019).

[40] This is not to say either that all biomedical interventions are unnatural or indeed that the emerging entity is unnatural. It is to say that the nature of that entity differs from our nature. There may be no bright lines separating us from this new being but that too is a reason for caution.

Bostrom, N. (2003). Human genetic enhancements: A transhumanist perspective. *Journal of Value Inquiry*, 493–506.

Bublitz, J., & Merkel, R. (2009). Autonomy and authenticity of enhanced personality traits. *Bioethics*, 360–374.

Buchanan, A. (2011). *Beyond humanity?: The ethics of biomedical enhancement*. Oxford and New York: Oxford University Press.

Carter, J., & Pritchard, D. (2019). The epistemology of cognitive enhancement. *The Journal of Medicine and Philosophy*, 220–242.

Caruso, G. (2020). *Free will and consciousness*. Lexington Books.

Cushman, F., Young, L., & Greene, D. (2010). Multi-sytem moral psychology. In *The moral psychology handbook* (pp. 47–71). Oxford: Oxford University Press.

Douglas, T. (2008). Moral enhancement. *Journal of Applied Philosophy*, 228–245.

Duff, R. A. (1993). Choice, character and criminal liability. *Law and Philosophy*, 345–383.

Fitzgerald, K. (2008). Medical enhancement: A destination of technological, not human, betterment. In B. Gordijin & R. Chadwick (Eds.), *Medical enhancement and post-modernity* (pp. 39–55). Springer.

Fukuyama, F. (2004). Transhumanism. *Foreign Policy*.

Harris, J. (2011). Moral enhancement and freedom. *Bioethics*, 102–111.

Jones, O., Montague, R., & Yaffe, G. (2020). Detecting mens Rea in the brain. *University of Pennsylvania Law Review*.

Jotterand, F. (2014). Questioning the Moral Enhancement Project. *American Journal of Bioethics*, 1–3.

Juengst, J. (2019). *Stanford encyclopedia of philosophy*. https://plato.stanford.edu/entries/enhancement/.

Kass, L. (2003). *Beyond therapy: Biotechnology and the pursuit of happiness*. President's Council on Bioethics, Executive Office of the President.

Korsgaard, C. (2008). *The constitution of agency: Essays on practical reason and moral psychology*. Oxford: Oxford University Press.

Melo-Martín, I. D. (2018). The trouble with moral enhancement. In L. Coyne & M. Hauskeller (Eds.), *Moral enhancement: Critical perspectives—Royal Institute of Philosophy Supplements* (pp. 19–33). Cambridge: Cambridge University Press.

Moore, M. (1979). Legal conceptions of mental illness. In B. Brody, & J. Engelhardt (Eds.), *Mental illness: Law and public policy, philosophy and medicine* (pp. 25–69).

Moore, M. (2015). The quest for a responsible responsibility test: Norwegian Insanity Law after Breivik. *Criminal Law and Philosophy*, 645–693.

Parens, E. (2005). *Authenticity and ambivalence: Toward understanding the enhancement debate* (The Hastings Center Report, 34).

Patterson, D., & Pardo, M. (2013). *Minds, brains and law*. Oxford University Press.

Persson, I., & Savulescu, J. (2008). The Perils of cognitive enhancement and the urgent imperative to enhance the moral character of humanity. *Journal of Applied Philosophy*, 162–177.

Pugh, J. (2019). Moral bio-enhancement, freedom, value and the parity principle. *Topoi*, 73–86.

Putnam, H. (2012). *Philosophy in an age of science*. Harvard University Press.

(n.d.). R v M'Naghten (1843) 8 E.R. 718.

Racine, E., Dubljevic, V., Jox, R., Baertschi, B., Christensen, J., Farisco, M., … Muller, S. (2017). Can neuroscience contribute to practical ethics? A critical review and discussion of the methodological and translational challenges of the neuroscience of ethics. *Bioethics*, 328–337.

Sandel, M. (2007). *The price of perfection: Ethics in the age of genetic engineering*. Cambridge: Harvard University Press.

Savulescu, J., & Persson, I. (2012). Moral enhancement, freedom and the God machine. *The Monist*, 399–421.

Shaw, E. (2017). Moral worth, biomedical moral enhancement and communicative punishment. *Journal of Law Information and Science*.

Singer, P. (2005). Ethics and intuitions. *The Journal of Ethics*, 331–352.

Sparrow, R. (2014). Better living through chemistry? A reply to Savulescu and Persson on 'moral enhancement'. *Journal of Applied Philosophy*, 23–32.

Specker, J., Schermer, M., & Reiner, P. (2017). Public attitudes towards moral enhancement. *Neuroethics*, 405–417.

Bebhinn Donnelly-Lazarov As Professor of Neuroscience, Law and Legal Philosophy, Bebhinn's research interests lie in jurisprudence and criminal law theory. Her book on criminal attempts, published by Cambridge University Press in 2015, is structured around an Anscombian account of intentional action. Recent and ongoing work explores our understanding of the mind and considers its implications for criminal responsibility, *mens rea*, and for defences. Bebhinn has begun to write a book on law and consciousness.

Retributivism, Consequentialism, and the Role of Science

Andrea Lavazza and Flavia Corso

Introduction: Science and the Law

Why do we punish? And are we justified in punishing an offender? These questions, traditionally pertaining to philosophical thinking (moral philosophy and philosophy of law), are nowadays characterized by many scientific aspects and concern neuroscience as well. Recent neuroscientific findings related to brain mechanisms that are deemed responsible for various types of behaviour, including that of criminals, have introduced a deterministic perspective on human decision-making (Roskies, 2006). The theoretical implications of these findings have led, more and more often, to radically sceptical interpretations as regards the existence

A. Lavazza (✉)
Centro Universitario Internazionale, Arezzo, Italy

University of Pavia, Pavia, Italy

F. Corso
Università di Genova, Genoa, Italy

S. Ligthart et al. (eds.), *Neurolaw*, Palgrave Studies in Law, Neuroscience, and Human Behavior,
https://doi.org/10.1007/978-3-030-69277-3_11

of free will. According to these views, we are not actually free in our behaviour or, at least, we are heavily restricted in our decision-making process. These new perspectives, it has been argued (Greene & Cohen, 2004), call for a radical reform of criminal justice, since the concepts of free will and personal responsibility constitute the fundamental premises of penal liability, in line with the common sense belief that individuals are free and responsible agents who deserve punishment if they violate the established rules.

Starting from the experiments carried out by neuroscientist Benjamin Libet (Libet et al., 1983), the common sense belief in free will has been proven empirically inaccurate, although this finding is still the subject of heated debates (Lavazza, 2016). The well-known experiments conducted by Libet and colleagues showed how, contrary to what had been believed thus far, the awareness of one's decision to perform an action (which in the experiment was flexing of a finger or a wrist) occurs about a third of a second later than the beginning of a specific brain activity, known as "readiness potential". Subsequently, many scholars interpreted those findings as a demonstration of the illusory character of free will; in other words, it was assumed that the brain determines the conscious decision and the action taken and that, consequently, the basis and the cause of a supposedly conscious movement are actually unconscious brain mechanisms over which the agent cannot exercise any control.

This idea of "illusionism" concerning free will has been widened beyond this strict empirical experiment. The ancient idea of determinism has recently merged with new philosophical ideas and psychological findings related, for example, to so-called situationism (cf. Bowers, 1973; Ross & Nisbett, 1991; Lavazza, 2019). This theory focuses specifically on the influence that external circumstances exert on the agent's decision-making process, suggesting that situational factors, rather than character traits, are what determines one's behaviour. For example, individuals who unexpectedly find a coin on their way home are more well-disposed towards a person asking for a favour than other control individuals, the coin being the only factor distinguishing their situations (Isen & Levin, 1972). In this theoretical framework, the agent is not able to exercise the necessary control over their actions, which allows to have responsibility attributed to them—in fact, the agent is supposed to have no real control over their own decisions and actions. In this view, human behaviour is guided by processes that bypass the conscious will of individuals, because actions do not causally descend from conscious deliberation, but from processes and factors over which individuals have no control (Pereboom, 2001).

Now, if we admit that there is no causal link between the actions we perform and our decision to carry them out, this assumption threatens the retributive model of justice on which punishment in the criminal justice system is based. If we are not free in the sense in which we commonly understand this term, or in the sense necessary in order to have real moral responsibility in the basic desert sense, how can we be subject to justified approval or disapproval? In addition, the development of genetics has made it possible to understand the influence that specific alleles can have on behaviour, particularly on anti-social and violent conducts (Raine, 2013). This does not amount to some new kind of determinism, because no variant of a single gene involves the certainty that the individual who carries it will invariably behave in a certain way. However, the interaction between genes and the environment can tell us a lot about people's behavioural tendencies and seems to be able to further reduce the scope of freedom of the subject.

All these aspects taken together have led to reconsider science's contribution to the law. And in the last two decades, a number of scholars, mainly with medical or neuroscientific training, have proposed radical reforms of criminal law in order to make it coherent with new knowledge on human behaviour, its mechanisms, and causes (cf. Greene & Cohen, 2004). In general, there has been a strong cultural trend in recent years towards the naturalization of moral and normative concepts based on the success of science and its growing role in our lives. However, the debate is still unsettled, and the positions are polarized. On the one hand, important reforms have been carried out, e.g. concerning juvenile criminal law in the Netherlands in 2014: on the basis of neuroscientific results showing the incomplete maturation of the brain of adolescents, it was decided that they should be judged less severely for crimes of impulse (Barendregt & Van der Laan, 2019; Schleim, 2019). On the other hand, there are many scholars who continue to deny the impact of neuroscience on the law (cf. Bigenwald & Chambon, 2019).

From this scientific perspective, when it comes to punishment, a shift from retributivism to consequentialism has been proposed (Greene & Cohen, 2004; Pereboom, 2001; Focquaert et al., 2019). Retributivism—the idea that wrongdoers deserve a penalty insofar as they have knowingly done wrong and punishment is deemed as a valuable end itself—is the default position of the common sense view and of criminal systems, even though most contemporary Western criminal systems can be defined as "mixed", that is, based on a retributivist ground but mitigated by consequentialist elements. Consequentialism, instead, is not directly or primarily interested in what happened in the past, but rather in future effects of the actions that are carried out after the violation has been committed. The consequentialists intend punishment to be for instance a deterrent so as to reduce further crime with the aim of protecting society and rehabilitate the offender.

When considering scientific data, it becomes challenging to rationally justify retributivism. Pure retributivism is linked to the idea that punishment is the only way to account for the conception of autonomous moral subjects, so that the punishment is nothing other than the consequence of the free choice of an individual. When asked "why do we punish?", retributivism states that punishment finds its reasons in itself, as a restitution of the evil committed by an agent who has free will. But if neuroscientific findings are properly taken into account, there does not seem to be a valid reason for punishment to be considered right in itself, since the agent is not actually responsible for their actions—they are not truly free to do otherwise.

For this reason, more detailed knowledge on the functioning of the brain pushes towards the preventative function of the punishment advocated by consequentialism. In this view, the criminal offender is no longer intended as an autonomous individual who is responsible for their actions, but rather as a dangerous person that must be subjected to care and rehabilitation. To the question "why do we punish?", consequentialism can only answer by looking forward, considering the punishment in its function of prevention of further crimes.

But things are more complex than they may seem. First of all, the science-based justification of consequentialism, i.e. free will illusionism, is far from being well corroborated (Lavazza, 2019). Secondly, if we want to take the path of naturalization, both retributivism and consequentialism are plausible candidates for this procedure. We are therefore faced with two pragmatically contradictory lines of naturalization of criminal justice, namely that indicated by cognitive neuroscience and that which refers to the naturalistic explanation of the origin of morality and law, conceived as the result of the dynamic adaptation of the human species to its natural and social environment. In this sense, our criminal system may also be explained from an evolutionary perspective (Lavazza & Sammicheli, 2012).

We cannot dwell here on what it means to naturalize a research field and, within it, a phenomenon or concept (cf. Kitcher, 1992; Petitot, 1999; Putnam, 1990). Suffice it to say that there are two main approaches in this respect. One is a more methodological approach, *à la* Quine, whereby, for example, epistemology should be transformed into a branch of descriptive psychology and the normative notion of justification should be replaced by a naturalist explanation of the link between sensory inputs and theoretical hypotheses, i.e. between observations and inferences. Another approach is ontological, so that the physicalist requirement is decisive: for example, in order to be the object of study of a natural science (characterized by laws or generalizations of nominal scope), mental states must be physical, because natural sciences recognize citizenship only to physical objects, events, and causal links (Carnap, 1931; Neurath, 1931).

Now, according to a general naturalistic paradigm, both consequentialism and retributivism, as practices and as justificatory frameworks of punishment, respond to the reality of the facts: if it is true that there is no valid reason to consider the punishment of criminals as just in itself, since they have no real control over their decision-making processes, it is equally true that the natural tendency to punish criminals has been crystallized in the course of evolution as a more functional adaptive strategy for the survival of groups and societies, and this consideration would make it difficult to abandon retributivism, also in the light of other philosophical considerations, which we will look at later.

So far we have seen how neuroscience can explain why, on the behavioural level, we are prone to punish offenders—an instinct which may lead to accept retributivism as a theory of justification of the punishment—and why we have also good reasons not to, since if people are genetically predisposed to deviance, there is no room for the idea of moral desert. So, how can we lean towards one or the other justification of punishment if both appear to be empirically plausible and grounded? This is the main issue we intend to address here. Therefore, we will analyse both lines of naturalization of consequentialism and retributivism, highlighting their strengths and weaknesses.

Consequentialism Naturalized

Many researchers in favour of a consequentialist naturalization of criminal justice have based their arguments on the findings of cognitive neuroscience, which—despite not being subject to a single interpretation—seem to indicate that free will is an illusion, or at least that its importance should be drastically reduced (Smilansky, 2000; Cashmore, 2010; Wegner, 2002; Harris, 2012). From this, it follows that moral responsibility, which theoretically requires personal freedom, is also a fallacious concept. The criminal act is therefore determined by a series of brain mechanisms and genetic make-ups over which the individual is not able to exercise full control, so that the inability to do otherwise than what their brain functioning imposes makes them immune to any classical imputation of justice.

For these reasons, neuroscience would lean towards alternative theories of punishment, which share the principle that punishment finds its justification in its preventive function. The punishment, in this case, is not fair in itself but only in relation to the beneficial effects that it produces on a social—or personal—level.

It is thus understood that the legitimacy of punishment derives from the right of society to self-defence as a means of guaranteeing order and social well-being. The gaze of justice is therefore turned towards the future and reflects the maxim *punitur ne peccetur*, and the ethical model to which these conceptions of punishment refer is of a teleological-consequentialist kind (Bentham, 1789). In other words, retributivism is an "unfair" criminal model according to a (neuro)scientific view, because there is no one that deserves punishment in the basic sense. Here, it can be noticed that this kind of consequentialism makes use of a moral argument about unfairness in addition to scientific data. The argument of unfairness seems anchored to clear and shared moral intuitions, but from a scientific-naturalistic point of view, resorting to shared intuitions is not uncontroversial.

Scientifically based consequentialism claims credit for not considering punishment just in itself and for moving away from a vindictive perspective of justice, emphasizing the humanitarian aspects of punishment. In its essence, consequentialism adopts a humanitarian theory of punishment based on the idea that punishment should not be justified by the concept of desert but should be placed in the preventive perspective of the protection of collective well-being. Consequentialism proposes a change in the consideration and perception of the offender, understood as a person with should be morally and socially re-educated (also so that he is no longer a danger) and not punished for something he committed in the past. In this vein, consequentialist punishment tries to minimize the suffering of the wrongdoer and does not abuse the prison institution.

These features are even more crucial if we accept the interpretations given by some cognitive neuroscientists on determinism and free will illusion. Greene and Cohen (2004) have highlighted how the law has so far remained substantially deaf to the appeals of scientifically based consequentialism, despite the supposed truth of some kind of determinism that requires us to consider every human behaviour as the result of an external coercive force (cf. Pereboom, 2001, 2014). Greene and Cohen notoriously underlined this inconsistency between common sense and the criminal system by means of the Mr. Puppet mental experiment. If an individual hypothetically designed by neuroscientists to be antisocial (Mr. Puppet) commits a crime, we are naturally inclined not to hold him responsible; this would not happen if the individual had not been previously manipulated by scientists.

This thought experiment aims to demonstrate that if radical determinism is endorsed, consequentialism is the only acceptable theory of punishment. Although he is a dangerous subject for society, Mr. Puppet should be considered neither guilty nor punishable. In fact, nobody is guilty in the basic desert sense, if we consider the scientific picture of human behaviour. From a consequentialist standpoint, there is no substantial difference between Mr. Puppet and all of us, as we are all puppets at the mercy of influences of various kinds (genes, environment, society, culture). In this vein, moral responsibility presupposes a free will that individuals do not actually have, which is why accepting determinism entails an overturning of the law in favour of a consequentialist view.

Leaving aside the factual and theoretical premises of this approach—whose validity remains objectionable (Mele, 2014)—opting for a preventative justice system means taking for granted a univocal definition of "positive effects" and tackling the thorny problem of the legitimate means to achieve them. Consequentialism has the ultimate goal of the reduction of crime (as well as the maximization of social well-being), but the means employed to obtain this goal can be often controversial as they are utilitarian in nature, i.e. oriented to the aim and not concerned about the individual rights of the people involved.

In the name of the common good, consequentialism justifies the instrumental treatment of individuals to achieve this goal, ignoring the Kantian principle by which every person should be considered an end in itself. Thus, consequentialism may not hesitate to sacrifice, for example, the right to privacy of one's brain (Farah, 2005; Farah et al., 2009), the right to mental self-determination (Ryberg, 2012; Bublitz & Merkel, 2014), the freedom of movement and association, on the basis of an arbitrary hierarchy of moral norms. Some researchers have already raised important ethical issues related to the application of neuroimaging techniques: the main problem lies in the identification of the right balance between individual rights, such as brain privacy and safeguarding social well-being.

The discovery of the efficacy of serotonergic drugs for reducing aggression (Coccaro & Kavoussi, 1997; Cherek et al., 2002) has also stirred many ethical concerns. Consequentialism shares the premise that criminals who are potentially harmful to society may be medically treated as they cannot control their own behaviour. However, is it ethically acceptable to require criminals to take SSRI drugs in exchange for early release? Ryberg (2012) objects that we have three reasons to be sceptical about such drug therapies. First, they threaten the right to authenticity, as drugs change the individual's personality (the concept of authenticity, however, has no clear and shared definition; also, contrary to what Ryberg claims, pharmacological therapies in many cases allow the person to be who they really are, so to speak, only without the burden of suffering due to mental illness [Kraemer, 2011]). Second, the chemical manipulation of the brain prevents the person from accessing their self-knowledge, which is necessary for understanding the triggers of mental problems (even if self-knowledge presupposes at least mental clarity, which cannot be given if the person is altered by the disease). Third, there is an element of coercion in a context where drug therapy is offered as an alternative to a prolonged prison sentence (as this alternative does not put the subject in the condition of being truly free to choose).

Furthermore, consequentialism could also have strong repercussions on reproductive rights. Although provocatively, LaFollette (1980) and Raine (2013) proposed parental licensing as a method to allow only the most competent parents to have children. According to these researchers, this would prevent both domestic violence and the formation of potential future criminals due to inadequate parents. In this view, the procedure for issuing the authorization to procreate is comparable to that necessary to be able to drive a car: if one does not have the necessary skills to carry out a certain activity, which can prove to be potentially dangerous, then they do not have the right to undertake it. However, this particularly radical proposal could favour a biopolitical model in contrast with the right to reproduction and parenthood, which is generally considered an absolute and inalienable human right.

It can also be objected, even within a consequentialist logic, that the theoretical basis of consequentialism can cause some unintended and disadvantageous effects as well. Experiments conducted by Vohs and Schooler (2008) have shown that the belief in determinism increases the feeling of irresponsibility and predisposes people to deception. On the contrary, the feeling of being free to choose empowers the responsibility of people, leading them to adopt prosocial behaviour (Baumeister et al., 2009). In this sense, retributivism could produce more positive effects than those that consequentialism tries to achieve with merely a forward-looking approach based on a scientific view of human motivations and behaviour.

Finally, another problem is related to the definition of "potential criminal" (Stearns, 1919), literally people strongly inclined to commit crimes both due to some mental disorders or voluntary disrespect for others. The prevention of crime, based on accurate and objective empirical assessments, seems unlikely and, in some cases, morally unacceptable. In particular, it is argued that, to prevent deviance, society could make use of new neurotechnologies and scientific discoveries. However, these would inevitably clash with the sphere of individual rights and freedoms (Douglas, 2014). For the well-being of the community, pure consequentialism could paradoxically justify the punishment of the innocent: if punishment is justified on the basis of its positive consequences, and one of these positive consequences is deterrence as a preventive strategy, then it follows that the punishment of the innocent can be justified in a consequentialist perspective.

The quarantine model suggested by Pereboom (2014) and then taken up by Caruso (2016) does not endorse a full-fledged consequentialism but can be deemed as a cognate position opposite to retributivism. In their view, the preventive detention of those who prove to be socially dangerous constitutes a more humane alternative to the retributive theory of punishment. Following a reasoning similar to that by which we quarantine infected subjects in the name of public health, the preventive confinement of potential criminals is supposed to bring benefits in terms of public safety. By abandoning the concept of punishment and accepting the ideas of re-education and rehabilitation, quarantine would therefore have the advantage of protecting both society and the well-being of (potential) wrongdoers.

However, many objections have been raised against this model. One could object that there is a lack of differentiation between the various types of criminals (Corrado, 2016), so that, for example, sexual violence and small thefts, crimes perpetrated by healthy subjects and crimes committed by individuals with partial or total mental disorders are all put on the same level. Crime and pathology would become essentially the same thing, and therefore, there would be no reason to make any kind of distinction between them. In addition to being very counterintuitive from the point of view of common sense, the quarantine model could also actually favour crime, facilitating non-recursive crimes on the part of one-time offenders not otherwise dangerous (Smilansky, 2017). Given the quarantine mode does not imply punishment but only to be comfortably confined for a period of time (especially if you are deemed a non-dangerous offender for the future), it seems legitimate to speculate that some offenders may be motivated to commit a single crime (the murder of the violent father) that they would not commit in a retributivist scenario (Lavazza et al., 2020).

A different criticism raised against the quarantine model refers to the idea of genetic justice. In particular, some thinkers have pointed out the paradox of how interned individuals would be the ones most in need of state aid. Their socially disadvantageous genetic heritage is in fact the outcome of the natural lottery, and the state should try to reduce the consequences deriving from genetic differences. With the quarantine model, however, internees may suffer a double injustice, a natural one (the defective genetic heritage) and a social one (isolation). The quarantine model, therefore, does not attenuate, but indeed amplifies, genetic injustice (Lavazza et al., 2020).

Retributivism Naturalized

If we consider retributivism from a naturalized point of view, it seems to be as legitimate and plausible as consequentialism, both in terms of justification and on the pragmatic-social level. Indeed, as mentioned, the second way of naturalizing the approach to criminal justice refers to a reconstruction of the moral and juridical phenomenon in an evolutionary key. According to this view, which is based on Darwinian evolutionism, moral norms are a product of human evolution. Our current criminal practices are therefore governed by an internal evolutionary dynamic that continues to proceed by trial and error, according to the logic of natural selection. In other words, the natural drive to punish the transgressors of the social order constitutes the most advantageous adaptive strategy in evolutionary terms. Accordingly, it would be a mistake to give up retributivism. One may also argue, for instance, that consequentialism is only apparently immune to retributive tendencies, whereas it justifies imposing neurointerventions to criminal offenders.

This kind of treatment may as well align with retributivism because it may symbolize the infliction of a penalty (Ryberg, 2018). Therefore, it seems that retributive drives have insinuated very deeply in our psychological features.

This evolutionary naturalization justifies adherence to the theories of punishment endorsed by retributivism. These kinds of theories refer to the maxim *punitur quia peccatum est*: both our practices and the law mainly focus on the past, on a committed injustice which, in order to be compensated for, requires the punishment of the wrongdoer and the restitution of the evil committed with a criminal offense. In other words, the punishment, if proportionate to the seriousness of the crime, finds its justification in itself, as it serves to restore a violated balance in society.

Many studies seem to show the existence of this natural drive for cooperation: people naturally tend to be kind to those who are kind and to punish those who do not comply to the rules, regardless of the consequences. Fehr and Gächter (2000, 2002) have demonstrated a spontaneous inclination to punish "free riders", even when the punishment is very expensive or does not bring any personal advantage to the punisher. In the absence of the institutions responsible for imposing sanctions on offenders, individuals are willing to punish wrongdoers even when punishment requires a high personal cost (Hauert et al., 2007). Similarly, the experiments conducted by Boyd and colleagues (2003) have shown that people tend to punish the wrongdoers even if punishment involves more costs than benefits for the punisher.

This instinct to punish offenders selflessly reinforces social norms (Fehr & Fishbacher, 2004) and is supposedly transmitted by the conformist imitation of the most frequent social behaviour within the population (Henrich & Boyd, 2001). Although costly punishment does not bring benefits to the individual punisher in the short term, it is likely that it will positively affect the social group as a whole in the long-term perspective (Gächter et al., 2008). Cooperation has proved to be the most successful survival strategy for our species, and this selfless instinct to punish offenders reinforces social norms oriented to cooperation. The fitness of groups capable of cooperation and not fraught with free riding is generally higher than that of other groups whose members are less prone to cooperate. In this vein, punishing behaviour against people who are not cooperating with the group proves to be highly adaptive by promoting the success of the group in its environment, both natural and social (Fehr & Fishbacher, 2004). It can sound a consequentialist approach, but we need to distinguish a potential consequentialist reason for the rising of retributivist drives and the content of the consequentialist view of punishment. The former is the reason why people "learned" to punish every time one is not complying with the rules, the latter is the idea that punishment is oriented and instrumental to obtain future goals.

In other words, one could affirm that retributivism is deeply rooted in human mind/brain as the result of a decisive force in the evolution of human cooperation (De Quervain et al., 2004), constituting an adaptive behavioural trait that brings benefits to the social organism in the long term (Trivers, 1971). According to this perspective, those who place themselves in a strong condition of reciprocity tend to cooperate with others and to punish, regardless of the costs of the punishment, those who do not cooperate (Gintis, 2000), and this behaviour would find an explanation in terms of evolution of the social body.

The socio-evolutionary origin of moralistic punishment seems to be confirmed in experiments that demonstrate how, as the audience increases, there is also an increase in the tendency to punish wrongdoers (Kurzban et al., 2007). Retributivism seems therefore to derive from this ancestral drive which proved to be the most suitable and functional means for safeguarding the survival and the flourishing of the social group.

As Lavazza and Sammicheli (2012) wrote:

> The law therefore is not a hypostasis found in the heavens like the Platonic essences [...]. It is instead a complex product of evolution, by which some social behaviors have proven beneficial for our individual survival and for our social prosperity. (2012, p. 214)

The cooperative nature of human beings could also be seen in their morphological characteristics. Kobayashi and Kohshima (2001) proposed the Cooperative Eye Hypothesis, according to which the morphological characteristics of the human eye seem to favour the monitoring of the direction of the gaze, a capacity that proves useful in contexts of cooperation and communication between members of the same species and which occurs in the neonatal age (Tomasello et al., 2007). Furthermore, some experiments have shown that people tend to adopt prosocial behaviours even under a stylized gaze; in short, the idea of being observed could be enough for human beings to engage with socially appropriate behaviour, namely rules-abiding and cooperative behaviour (Haley & Fessler, 2005). These studies highlight the substantially social and empathic nature of human beings (Hoffman, 2004), their ancestral need to communicate and cooperate with other members of the group and the drive to prevent others from violating the commitments to the group by punishing free riders and wrongdoers.

The discovery of mirror neurons (Rizzolatti & Sinigaglia, 2007), which constitute a component of the biological basis of empathy, is another element in favour of this evolutionary interpretation of human cooperation. It is therefore plausible that, at a certain stage of evolution, human beings "understood" that cooperation was particularly helpful for surviving in dangerous contexts:

> [...] we came to have these "moral sentiments" because our ancestors lived in environments, both natural and socially constructed, in which groups of individuals who were predisposed to cooperate and uphold ethical norms tended to survive and expand relative to other groups, thereby allowing these prosocial motivations to proliferate. (Bowles & Gintis, 2011, p. 1)

We could say that we cooperate for the simple "pleasure" of cooperating (which we have unconsciously learned during evolution), even when cooperation does not produce individual utility.

Given these considerations, retributivism might not need to be further justified on a rational level (Nichols, 2013). However, *prima facie* this evolutionary naturalization does not come without pitfalls. It can be observed that this approach risks running into the trap of the naturalistic fallacy (already highlighted by Hume, and then taken up by

Moore): deriving moral values and prescriptions from facts and observations constitutes faulty reasoning. One could accept that human beings have a natural drive to punish but leaping from the descriptive to the prescriptive level is not philosophically justifiable. If one accepted to infer what it ought to be (justice) from what it is (natural instincts), one would risk justifying any other behavioural tendency merely based on its adaptive function. The instinct to punish, originally adaptive in small groups and exercised by individuals, can easily result in an irrational tendency if inserted in a different social context, for example, in larger groups, where the infliction of punishment is institutionalized (Lavazza & Sammicheli, 2012). We also have other drives, such as envy, which however we are not so prone to accept or encourage. In short, it is not sufficient that a given behaviour is determined by an instinct for it to be morally justified.

From a naturalized viewpoint, we should also consider that our reference to a moral framework, based essentially on intuition or shared values, could prove not so robust. Indeed, take the so-called evolutionary debunking argument proposed by Street (2006). In this view, human systems of moral evaluative judgements are "thoroughly saturated with evolutionary influence" because natural selection has shaped human psychological dispositions. Evolution has selected those moral evaluative judgements according to biological fitness (rather than based on some moral truths of the realist kind). If human moral beliefs, shaped by evolution, aligned with moral truths, then this would be sheer coincidence. We are not justified in assuming that such a coincidence has occurred. So, we cannot justifiably believe that our moral beliefs accurately represent independent moral truths. Moral realism should therefore give way to moral scepticism.

Criticism has been raised against this argument (Carruthers & James 2008; Wielenberg, 2010), which we will not go into here. What we do want to point out is that the evolutionary debunking argument tells us that our moral values can be taken to be relative because they are shaped by evolution, which proceeds without a design or a purpose. Therefore, it does not make sense to say that we should not derive moral values from facts, since all values and prescriptions are dependent on facts related to evolution. And the values we have from evolution seem to be the fittest in terms of cooperation and survival of human groups. So, naturalized retributivism cannot be objected to in the terms exposed above and seems to be perfectly plausible from an evolutionary point of view.

Nevertheless, we can state that retributivism pays insufficient attention to different types of crime. According to pure retributivism, which is no longer considered in the contemporary penal systems, wrongdoers must be punished regardless of, for example, their mental health and the social environment in which they have grown up (Moore, 1997). Another pitfall of retributivism is its insistence on harming the wrongdoer, which today comes in the privation of freedom (by imprisonment) and wealth (through fines). So, retributivism is deemed as less humane than other criminal approaches. It should be noted that the conditions of the detainees often further exacerbate frustration and aggression in subjects who are easily sensitive to them. Although very controversial, the "Lucifer effect" resulting from the well-known experiment carried out at Stanford University by Zimbardo (2008) in 1971 may demonstrate how the context and power of institutions can also influence the development of aggression.[1]

It has been argued that the view underlying retributivism does not sufficiently take into account the well-being of the punished person. Nonetheless, one may maintain that contrary to consequentialism retributivism follows the Kantian principle according to which one should take the person *as an end in itself*. But this is true only if the criminal is understood as a free subject capable of acting in accordance with their nature (which can be good or evil), whereas according to neuroscientific form of consequentialism, the individual is unable to act freely. For this reason, the absence of rehabilitative and re-educational aspects in punishment contributes to fuelling the social stigma (Goffman, 1963) towards socially unwelcome individuals, favouring the isolation of the criminal from society even after having completed their sentence. Both retributivism and consequentialism seem to require some form of confinement or isolation to protect society, but it is one thing to be imprisoned because one "deserves" it and another thing to be isolated to prevent one from harming others; it is one thing to suffer as punishment for one's chosen behaviour and another thing to be compulsorily

[1] For criticism about this experiment, see Le Texier (2019).

cured for a disease which made us offend others. The high recidivism rate of former prisoners in penal systems which endorse retributivism seems to confirm that prison as a penalty does not contribute to the rehabilitation of the person. For this reason, the imposition of retributive penalties, which are harmful in nature, turns out to be not fully effective in reducing crime (Golash, 2005).

Conclusion: The Middle Way of Neurolaw

Now, neuroscientific findings that seem to downsize the weight of freedom and personal responsibility might tilt towards a reform of the criminal system, at least as concerns the punishing process. Since we take seriously both the findings of brain science and the evolutionary naturalization of moral norms and values, we have tried to consider the two classic approaches of punishment in this perspective.

However, on a pragmatic level, we are faced with a stalemate: on the one hand, the discovery of genes and brain malfunctions that are responsible for deviant conduct leads to a naturalization that highlights the inconsistency of retributivist punishment, arguing in favour of a pure consequentialist approach. On the other hand, the punishment of the wrongdoer is a basic human drive; it has been embedded in human psychology as an adaptation which increases the fitness of the group since it has proven to be useful for social cooperation. In this vein, retributivism has a strong pragmatic value for the proper functioning of human societies, as is also shown by the fact that many people believe that wrongdoers, especially those who commit more serious crimes, should be punished with at least some afflictive measures (typically, a period of detention in prison). Although an objection can be raised towards retributivism about its moral justification, if we take the naturalization of ethics seriously, this argument loses a lot of its cogency.

So, we are confronted with divergent strands of naturalization with regards to two different but coexisting practices and approaches to punishment. The naturalistic paradigm explains both attitudes but develops a pragmatic contradiction: the criminal can be treated, so to speak, either as a "sinner" or as a sick person (Sapolsky, 2004). Although a form of confinement or isolation of offenders seems to be necessary in order to protect the well-being of the community, it can be said that a perspective shift occurs when it comes to providing justification for such a limitation of personal freedom; one might argue that such an external imposition might be understood as something that offenders simply "deserve" or as something "needed" to be done for reasons of public security.

It can be argued that neuroscience-based consequentialism (a specific form of the latter), by rejecting or devaluating the idea of personal freedom and responsibility, risks conflicting with the common sense psychology shaped by our evolution. Indeed, individuals generally have the feeling of being free to choose and liable for their actions, although that belief seems to be disconfirmed by recent neuroscientific findings (Monroe & Malle, 2010). In a sense, societies are believed to be capable of working only if they assume the existence of moral responsibility, even though the latter does not have clear and precise borders but comes in degrees within a continuum. Consequentialism also presupposes a questionable hierarchy between individual and collective rights. For the good of society, it runs the risk of overshadowing the importance of individual rights. Accordingly, on the pragmatic level, the consequentialist-like perspective on punishment—for example the quarantine model considered above—cannot be labelled per se as more "humane" than retributivism (Lavazza et al., 2021).

It therefore seems reasonable to suppose that a scientifically informed theory of criminal punishment cannot completely ignore either consequentialist or retributivist aspects. The former considers recent findings on brain functioning and the humanitarian view of punishment, which should not be afflictive per se; the latter takes into account a very relevant aspect of personal psychological motivations and social practices. Both are plausible naturalized approaches to criminal punishment.

If we take science seriously, we cannot ignore its findings and its effect on the law and, specifically, on the way we punish wrongdoers. As seen, we have good reasons to hold both approaches theoretically plausible and pragmatically helpful. We are social animals, and as such we tend to punish those who break the social rules in order to preserve cooperation and the fitness of the group; retributivism, in this perspective, is the most coherent legal theory. However, deviant behaviours can be strongly affected and conditioned by brain malfunctions such as to invalidate the liability of the offender, making the afflictive function of punishment unjust. And recent neuroscientific findings tell us that we are probably not free at all, so consequentialism turns out to be the best criminal approach to punishment.

The reconciliation of these two approaches in the light of naturalization of criminal systems is problematic from a pragmatic point of view, but neurolaw as an interdisciplinary endeavour can help address this challenge. Specifically, we should resort more frequently to neuroscience in courtrooms in order to assess how free and sane an offender is, thus making our criminal system more humane and less unnecessarily afflictive. But it seems that we cannot easily give up our retributivist intuitions as they are deeply rooted in our natural history. In fact, it is reasonable to say that only through a plurality of complementary approaches will it be possible to find a solution, albeit partial and not definitive, to the problem of who, how and for what reason should be punished.

References

Barendregt, C. S., & van der Laan, A. M. (2019). Neuroscientific insights and the Dutch adolescent criminal law: A brief report. *Journal of Criminal Justice, 65,* 3. https://doi.org/10.1016/j.jcrimjus.2018.05.010.

Baumeister, R. F., Masicampo, E. J., & DeWall, C. N. (2009). Prosocial benefits of feeling free: Disbelief in free will increases aggression and reduces helpfulness. *Personality and Social Psychology Bulletin, 35*(2), 260–268. https://doi.org/10.1177/0146167208327217.

Bigenwald, A., & Chambon, V. (2019). Criminal responsibility and neuroscience: No revolution yet. *Frontiers in Psychology, 10,* 1406. https://doi.org/10.3389/fpsyg.2019.01406.

Bowers, K. S. (1973). Situationism in psychology: An analysis and a critique. *Psychological Review, 80*(5), 307–336. https://doi.org/10.1037/h0035592.

Bowles, S., & Gintis, H. (2011). *A cooperative species: Human reciprocity and its evolution.* Princeton, NJ: Princeton University Press.

Boyd, R., Gintis, H., Bowles, S., & Richerson, P. J. (2003). The evolution of altruistic punishment. *Proceedings of the National Academy of Sciences, 100*(6), 3531–3535. https://doi.org/10.1073/pnas.0630443100.

Bublitz, J. C., & Merkel, R. (2014). Crimes against minds: On mental manipulations, harms and a human right to mental self-determination. *Criminal Law and Philosophy, 8*(1), 51–77. https://doi.org/10.1007/s11572-012-9172-y.

Carnap, R. (1931). Die physikalische Sprache als Universalsprache der Wissenschaft. *Erkenntnis, 2*(1), 432–465. https://doi.org/10.1007/BF02028172.

Carruthers, P., & James, S. M. (2008). Evolution and the possibility of moral realism. *Philosophy and Phenomenological Research, 77*(1), 237–244.

Caruso, G. D. (2016). Free will skepticism and criminal behavior: A public health-quarantine model. *Southwest Philosophy Review, 32*(1), 25–48. https://doi.org/10.5840/swphilreview20163214.

Cashmore, A. R. (2010). The Lucretian swerve: The biological basis of human behavior and the criminal justice system. *Proceedings of the National Academy of Sciences, 107*(10), 4499–4504. https://doi.org/10.1073/pnas.0915161107.

Cherek, D. R., Lane, S. D., Pietras, C. J., & Steinberg, J. L. (2002). Effects of chronic paroxetine administration on measures of aggressive and impulsive responses of adult males with a history of conduct disorder. *Psychopharmacology (Berl), 159*(3), 266–274. https://doi.org/10.1007/s002130100915.

Coccaro, E. F., & Kavoussi, R. J. (1997). Fluoxetine and impulsive aggressive behavior in personality-disordered subjects. *Archives of General Psychiatry, 54*(12), 1081–1088. https://doi.org/10.1001/archpsyc.1997.01830240035005.

Corrado, M. L. (2016). *Chapter one: Two models of criminal justice* (UNC Legal Studies, Research Paper No. 2757078). https://doi.org/10.2139/ssrn.2757078.

De Quervain, D. J., Fischbacher, U., Treyer, V., & Schellhammer, M. (2004). The neural basis of altruistic punishment. *Science, 305*(5688), 1254–1258. https://doi.org/10.1126/science.1100735.

Douglas, T. (2014). Criminal rehabilitation through medical intervention: Moral liability and the right to bodily integrity. *The Journal of Ethics, 18*(2), 101–122. https://doi.org/10.1007/s10892-014-9161-6.

Farah, M. J. (2005). Neuroethics: The practical and the philosophical. *Trends in Cognitive Sciences, 9*(1), 34–40. https://doi.org/10.1016/j.tics.2004.12.001.

Farah, M. J., Smith, M. E., Gawuga, C., Lindsell, D., & Foster, D. (2009). Brain imaging and brain privacy: A realistic concern? *Journal of Cognitive Neuroscience, 21*(1), 119–127. https://doi.org/10.1162/jocn.2009.21010.

Fehr, E., & Fischbacher, U. (2004). Third-party punishment and social norms. *Evolution and Human Behavior, 25*(2), 63–87. https://doi.org/10.1016/S1090-5138(04)00005-4.

Fehr, E., & Gächter, S. (2000). Cooperation and punishment in public goods experiments. *American Economic Review, 90*(4), 980–994.

Fehr, E., & Gächter, S. (2002). Altruistic punishment in humans. *Nature, 415*(6868), 137–140. https://doi.org/10.1038/415137a.

Focquaert, F., Caruso, G., Shaw, E., & Pereboom, D. (2019). Justice without retribution: Interdisciplinary perspectives, stakeholder views and practical implications. *Neuroethics, 1,* 1–3. https://doi.org/10.1007/s12152-019-09413-8.

Gächter, S., Renner, E., & Sefton, M. (2008). The long-run benefits of punishment. *Science, 322*(5907), 1510. https://doi.org/10.1126/science.1164744.

Gintis, H. (2000). Strong reciprocity and human sociality. *Journal of Theoretical Biology, 206* (2), 169–179. https://doi.org/10.1006/jtbi.2000.2111.

Goffman, E. (1963). *Stigma: Notes on the management of spoiled identity.* Englewood Cliffs, NJ: Prentice-Hall.

Golash, D. (2005). *The case against punishment: Retribution, crime prevention, and the law.* New York: New York University Press.

Greene, J., & Cohen J. D. (2004). For the law, neuroscience changes nothing and everything. *Philosophical Transactions of the Royal Society of London – Series B: Biological Sciences, 359*(1451), 1775–1785. https://doi.org/10.1098/rstb.2004.1546.

Haley, K. J., & Fessler, D. M. T. (2005). Nobody's watching?: Subtle cues affect generosity in an anonymous economic game. *Evolution and Human Behavior, 26* (3), 245–256. https://doi.org/10.1016/j.evolhumbehav.2005.01.002.

Harris, S. (2012). *Free will.* New York: Simon and Schuster.

Hauert, C., Traulsen, A., Brandt, H., Nowak, M. A., & Sigmund, K. (2007). Via freedom to coercion: The emergence of costly punishment. *Science, 316* (5833), 1905–1907. https://doi.org/10.1126/science.1141588.

Henrich, J., & Boyd, R. (2001). Why people punish defectors: Weak conformist transmission can stabilize costly enforcement of norms in cooperative dilemmas. *Journal of Theoretical Biology, 208*(1), 79–89. https://doi.org/10.1006/jtbi.2000.2202.

Hoffman, M. B. (2004). The neuroeconomic path of the law. *Philosophical Transactions of the Royal Society of London. Series B: Biological Sciences, 359*(1451), 1667–1676. https://doi.org/10.1098/rstb.2004.1540.
Isen, A. M., & Levin, P. F. (1972). Effect of feeling good on helping: Cookies and kindness. *Journal of Personality and Social Psychology, 21*(3), 384. https://doi.org/10.1037/h0032317.
Kitcher, P. (1992). The naturalists return. *The Philosophical Review, 101*(1), 53–114. https://doi.org/10.2307/2185044.
Kobayashi, H., & Kohshima, S. (2001). Unique morphology of the human eye and its adaptive meaning: Comparative studies on external morphology of the primate eye. *Journal of Human Evolution, 40*(5), 419–435. https://doi.org/10.1006/jhev.2001.0468.
Kraemer, F. (2011). Authenticity anyone? The enhancement of emotions via neuro-psychopharmacology. *Neuroethics, 4*(1), 51–64. https://doi.org/10.1007/s12152-010-9075-3.
Kurzban, R., DeScioli, P., & O'Brien, E. (2007). Audience effects on moralistic punishment. *Evolution and Human Behavior, 28*(2), 75–84. https://doi.org/10.1016/j.evolhumbehav.2006.06.001.
LaFollette, H. (1980). Licensing parents. *Philosophy & Public Affairs, 9*(2), 182–197.
Lavazza, A. (2016). Free will and neuroscience: From explaining freedom away to new ways of operationalizing and measuring it. *Frontiers in Human Neuroscience, 10*(262). https://doi.org/10.3389/fnhum.2016.00262.
Lavazza, A. (2019). Why cognitive sciences do not prove that free will is an epiphenomenon. *Frontiers in Psychology, 10*(326). https://doi.org/10.3389/fpsyg.2019.00326.
Lavazza, A., Levin, S., & Farina, M. (2021). Not so humane: Why the quarantine model of incapacitation fails to be the best criminal policy. Submitted.
Lavazza, A., & Sammicheli, L. (2012). *Il delitto del cervello. La mente tra scienza e diritto*. Torino: Codice edizioni.
Le Texier, T. (2019). Debunking the Stanford prison experiment. *American Psychologist, 74*(7), 823–839. https://doi.org/10.1037/amp0000401.
Libet, B., Gleason, C., Wright, E., & Pearl, D. (1983). Time of conscious intention to act in relation to onset of cerebral activity (readiness-potential): The unconscious initiation of a freely voluntary act. *Brain, 106*(3), 623–642. https://doi.org/10.1007/978-1-4612-0355-1_15.
Mele, A. R. (2014). *Free: Why Science hasn't disproved free will*. Oxford: Oxford University Press.

Monroe, A. E., & Malle, B. F. (2010). From uncaused will to conscious choice: The need to study, not speculate about people's folk concept of free will. *Review of Philosophy and Psychology, 1,* 211–224. https://doi.org/10.1007/s13164-009-0010-7.

Moore, M. (1997). *Closet retributivism—Placing blame: A general theory of the criminal law*. Oxford: Clarendon Press.

Neurath, O. (1931). Soziologie und Physicalismus. *Erkenntnis, 2,* 5–6.

Nichols, S. (2013). Brute retributivism. In T. Nadelhoffer (Ed.), *The future of punishment*. Oxford: Oxford University Press.

Pereboom, D. (2001). *Living without free will*. Cambridge: Cambridge University Press.

Pereboom, D. (2014). *Free will, agency, and meaning in life*. Oxford: Oxford University Press.

Petitot, J. (1999). *Naturalizing phenomenology: Issues in contemporary phenomenology and cognitive Science*. Stanford, CA: Stanford University Press.

Putnam, H. (1990). Why reason can't be naturalized. In *Philosophy, mind, and cognitive inquiry* (pp. 283–303). Dordrecht: Springer. https://doi.org/10.1007/978-94-009-1882-5_11.

Raine, A. (2013). *The anatomy of violence: The biological roots of crime*. New York: Vintage Books.

Rizzolatti, G., & Sinigaglia, C. (2007). *Mirrors in the brain: How our minds share actions and emotions*. New York: Oxford University Press.

Roskies, A. (2006). Neuroscientific challenges to free will and responsibility. *Trends in Cognitive Sciences, 10*(9), 419–423. https://doi.org/10.1016/j.tics.2006.07.011.

Ross, L., & Nisbett, R. E. (1991). *The person and the situation: Perspectives of social psychology*. New York: McGraw-Hill.

Ryberg, J. (2012). Punishment, pharmacological treatment, and early release. *International Journal of Applied Philosophy, 26*(2), 231–244. https://doi.org/10.5840/ijap201226217.

Ryberg J. (2018). Neuroscientific treatment of criminals and penal theory. In *Treatment for crime: Philosophical essays on neurointerventions in criminal justice*. New York: Oxford University Press.

Sapolsky, R. M. (2004). The frontal cortex and the criminal justice system. *Philosophical Transactions-Royal Society of London: Biological Sciences, 359*(1451), 1787–1796. https://doi.org/10.1098/rstb.2004.1547.

Schleim, S. (2019). 'Neurorecht' in Nederland. *Algemeen Nederlands Tijdschrift Voor Wijsbegeerte, 111*(3), 379–404. https://doi.org/10.5117/ANTW2019.3.005.SCHL.
Smilansky, S. (2000). *Free will and illusion.* Oxford: Clarendon Press.
Smilansky, S. (2017). Pereboom on punishment—Funishment, innocence, motivation, and other difficulties. *Criminal Law and Philosophy, 11*(3), 591–603. https://doi.org/10.1007/s11572-016-9396-3.
Stearns, A. W. (1919). The detection of the potential criminal. *Journal of the American Institute of Criminal Law and Criminology, 9*(4), 514–519. https://www.jstor.org/stable/1134126.
Street, S. (2006). A Darwinian dilemma for realist theories of value. *Philosophical Studies, 127,* 109–166.
Tomasello, M., Hare, B., Lehmann, H., & Call, J. (2007). Reliance on head versus eyes in the gaze following of great apes and human infants: The cooperative eye hypothesis. *Journal of Human Evolution, 52*(3), 314–320. https://doi.org/10.1016/j.jhevol.2006.10.001.
Trivers, R. L. (1971). The evolution of reciprocal altruism. *The Quarterly Review of Biology, 46*(1), 35–57. https://doi.org/10.1086/406755.
Vohs, K. D., & Schooler, J. W. (2008). The value of believing in free will: Encouraging a belief in determinism increases cheating. *Psychological Science, 19*(1), 49–54. https://doi.org/10.1111/j.1467-9280.2008.02045.x.
Wegner, D. M. (2002). *The illusion of consciousness.* Cambridge: MIT Press.
Wielenberg, E. J. (2010). On the evolutionary debunking of morality. *Ethics, 120*(3), 441–464. https://doi.org/10.1086/652292.
Zimbardo, P. (2008). *The Lucifer effect: Understanding how good people turn evil.* New York: Random House.

Dr. Andrea Lavazza is a Senior Research Fellow in Neuroethics at Centro Universitario Internazionale, and adjunct Professor at the University of Pavia. He specializes in philosophy of mind and neuroethics. His main research is in the field of neuroethics. He has published papers on enhancement, memory manipulation, cognitive freedom, and human brain organoids. His interests are focused on moral philosophy, free will, and law at the intersection with cognitive sciences. He is working on naturalism and its relations with other kinds of causation and explanation in philosophy of mind and philosophical anthropology.

Dr. Flavia Corso holds a II level master degree from the University of Genoa.

She collaborates with Andrea Lavazza on a variety of neuroethical issues, mainly focusing on the role of neuroscience in the field of criminal justice. Recently, she contributed to an Italian volume on neuroethics, discussing the neuro-legal issue of responsibility in the age of neuroscience.

Index

S. Ligthart et al. (eds.), *Neurolaw*, Palgrave Studies in Law, Neuroscience, and Human Behavior,
https://doi.org/10.1007/978-3-030-69277-3

The manufacturer's authorised representative in the EU is Springer Nature Customer Service Centre GmbH, Europaplatz 3, 69115 Heidelberg, Germany. If you have any concerns regarding our products, please contact ProductSafety@springernature.com

Printed and bound by CPI Group (UK) Ltd, Croydon, CR0 4YY
30/09/2026
02225459-0007